Fatigue Performance and Life-Cycle Evaluation of Existing Bridges

Wei Zhang
Steve C.S. Cai

Fatigue Performance and Life-Cycle Prediction of Existing Bridges

Dynamic Effects from Combined Random Loads of Winds and Vehicles

LAP LAMBERT Academic Publishing

Impressum/Imprint (nur für Deutschland/only for Germany)
Bibliografische Information der Deutschen Nationalbibliothek: Die Deutsche Nationalbibliothek verzeichnet diese Publikation in der Deutschen Nationalbibliografie; detaillierte bibliografische Daten sind im Internet über http://dnb.d-nb.de abrufbar.
Alle in diesem Buch genannten Marken und Produktnamen unterliegen warenzeichen-, marken- oder patentrechtlichem Schutz bzw. sind Warenzeichen oder eingetragene Warenzeichen der jeweiligen Inhaber. Die Wiedergabe von Marken, Produktnamen, Gebrauchsnamen, Handelsnamen, Warenbezeichnungen u.s.w. in diesem Werk berechtigt auch ohne besondere Kennzeichnung nicht zu der Annahme, dass solche Namen im Sinne der Warenzeichen- und Markenschutzgesetzgebung als frei zu betrachten wären und daher von jedermann benutzt werden dürften.

Coverbild: www.ingimage.com

Verlag: LAP LAMBERT Academic Publishing GmbH & Co. KG
Heinrich-Böcking-Str. 6-8, 66121 Saarbrücken, Deutschland
Telefon +49 681 3720-310, Telefax +49 681 3720-3109
Email: info@lap-publishing.com

Approved by: Baton Rouge, Louisiana State University, Diss., 2012

Herstellung in Deutschland (siehe letzte Seite)
ISBN: 978-3-659-13538-5

Imprint (only for USA, GB)
Bibliographic information published by the Deutsche Nationalbibliothek: The Deutsche Nationalbibliothek lists this publication in the Deutsche Nationalbibliografie; detailed bibliographic data are available in the Internet at http://dnb.d-nb.de.
Any brand names and product names mentioned in this book are subject to trademark, brand or patent protection and are trademarks or registered trademarks of their respective holders. The use of brand names, product names, common names, trade names, product descriptions etc. even without a particular marking in this works is in no way to be construed to mean that such names may be regarded as unrestricted in respect of trademark and brand protection legislation and could thus be used by anyone.

Cover image: www.ingimage.com

Publisher: LAP LAMBERT Academic Publishing GmbH & Co. KG
Heinrich-Böcking-Str. 6-8, 66121 Saarbrücken, Germany
Phone +49 681 3720-310, Fax +49 681 3720-3109
Email: info@lap-publishing.com

Printed in the U.S.A.
Printed in the U.K. by (see last page)
ISBN: 978-3-659-13538-5

TABLE OF CONTENTS

LIST OF TABLES ·· 3

LIST OF FIGURES ·· 5

ABSTRACT ·· 8

CHAPTER 1 INTRODUCTION ··· 9
  1.1 Bridge Aerodynamics··· 9
  1.2 Vehicle-Bridge-Wind Interaction Dynamics ··· 11
  1.3 Bridge Fatigue ·· 12
  1.4 Overview of the Book ·· 14
  1.5 References ··· 15

CHAPTER 2 FATIGUE RELIABILITY ASSESSMENT FOR EXISTING BRIDGES
  CONSIDERING VEHICLE SPEED AND ROAD SURFACE CONDITIONS ············· 20
  2.1 Introduction ··· 20
  2.2 Vehicle-Bridge Dynamic System ·· 21
  2.3 Modeling of Progressive Deterioration for Road Surface ···························· 23
  2.4 Prototypes of Bridge And Vehicle ··· 24
  2.5 Fatigue Reliability Assessment ·· 30
  2.6 Results and Discussions ··· 33
  2.7 Conclusions ··· 40
  2.8 References ··· 41

CHAPTER 3 RELIABILITY BASED DYNAMIC AMPLIFICATION FACTOR ON STRESS
  RANGES FOR FATIGUE DESIGN OF EXISTING BRIDGES ···························· 45
  3.1 Introduction ··· 45
  3.2 Stress Range Acquisition·· 47
  3.3 Dynamic Amplification Factor on Stress Ranges (DAFS) ··························· 56
  3.4 Fatigue Life Estimation·· 64
  3.5 Conclusions ··· 68
  3.6 References ··· 69

CHAPTER 4 PROGRESSIVE FATIGUE RELIABILITY ASSESSMENT OF EXISTING
  BRIDGES BASED ON A NONLINEAR CONTINUOUS FATIGUE DAMAGE
  MODEL ··· 71
  4.1 Introduction ··· 71
  4.2 Generating Stress Range History ·· 72
  4.3 Fatigue Reliability Assessment ·· 75
  4.4 Numerical Example ··· 79
  4.5 Concluding Remarks··· 92
  4.6 References ··· 93

CHAPTER 5 FINITE ELEMENT MODELING OF BRIDGES WITH EQUIVALENT
ORTHOTROPIC MATERIAL METHOD FOR MULTI-SCALE DYNAMIC
LOADS ·· 96
  5.1 Introduction ·· 96
  5.2 Equivalent Orthotropic Material Modeling Method ··· 97
  5.3 Modeling of Vehicle-Bridge Dynamic System ·· 102
  5.4 Numerical Example ·· 104
  5.5 Concluding Remarks··· 116
  5.6 References ·· 117

CHAPTER 6 FATIGUE RELIABILITY ASSESSMENT FOR LONG-SPAN BRIDGES UNDER
COMBINED DYNAMIC LOADS FROM WINDS AND VEHICLES ························ 119
  6.1 Introduction·· 119
  6.2 Finite Element Modeling Scheme for Long-Span Bridges ····································· 120
  6.3 Vehicle-Bridge-Wind Dynamic System ··· 124
  6.4 Acquisition of Stress Cycle Blocks ·· 129
  6.5 Fatigue Reliability Assessment·· 134
  6.6 Selected Results ··· 136
  6.7 Concluding Remarks··· 139
  6.8 References ·· 139

CHAPTER 7 CONCLUSIONS AND FURTHER CONSIDERATIONS···························· 143
  7.1 Summary and Conclusion ··· 143
  7.2 Future Work ··· 146

LIST OF TABLES

Table 2-1 RRC values for road roughness classifications ·· 24

Table 2-2 Major parameters of vehicle (3 axles) ·· 27

Table 2-3 Vehicle speed ranges ·· 28

Table 2-4 Summary of LSF parameters ··· 32

Table 2-5 Chi-Square test for the revised equivalent stress range Sw threshold =3.45Mpa (0.5 ksi) ··· 36

Table 2-6 Chi-Square test for the revised equivalent stress range Sw threshold =13.8 Mpa (2.0 ksi) ··· 36

Table 2-7 Mean and Coefficient of Variation of Sw (threshold =3.45Mpa(0.5 ksi)) ·············· 36

Table 2-8 Fatigue reliability index and Fatigue life (threshold =3.45Mpa(0.5 ksi)) ·············· 38

Table 3-1 Major parameters of vehicle (2 axles) ··· 52

Table 3-2 Major parameters of vehicle (3 axles) ··· 54

Table 3-3 Major parameters of vehicle (5 axles) ··· 56

Table 3-4 Number of years for cases with different truck combinations ························· 64

Table 3-5 Number of years for cases with different annual traffic increase rates ·············· 64

Table 3-6 Summary of LSF parameters ··· 66

Table 4-1 RRC values for road roughness classifications ··· 77

Table 4-2 Major parameters of a vehicle (2 axles) ·· 80

Table 4-3 Major parameters of a vehicle (3 axles) ·· 81

Table 4-4 Major parameters of a vehicle (5 axles) ·· 82

Table 5-1 Material property for bottom plate of the short-span bridge ·························· 105

Table 5-2 Major parameters of vehicle (3 axles) ··· 106

Table 5-3 Comparisons of natural frequencies of the short span bridge ······················· 108

Table 5-4 Dynamic Properties of six main Modes for Donghai Bridge ························· 109

Table 5-5 Comparisons of revised equivalent stress ranges for short span bridge ·············· 115

Table 6-1 Natural frequencies of the six modes for Donghai Bridge ································ 124

Table 6-2 RRC values for road roughness classifications ···································· 130

Table 6-3 Vehicle speed ranges ·· 131

Table 6-4 Number of vehicles passing the bridge detail in one day ···················· 131

Table 6-5 Wind velocity and directions for Donghai Bridge ··························· 134

Table 6-6 Representative mean wind velocity and probability ························· 134

Table 6-7 Summary of LSF parameters ·· 136

Table 6-8 Stress ranges due to single vehicle passage ································· 136

Table 6-9 Stress ranges and number of stress cycles due to wind loads in one hour ············· 137

Table 6-10 Stress ranges and stress cycles due to dynamic loads in one day ·················· 137

Table 6-11 Estimated fatigue life (year) ··· 138

LIST OF FIGURES

Fig. 2-1 Typical section of bridge (unit= meter) ⋯⋯⋯⋯⋯⋯⋯⋯⋯⋯⋯⋯⋯ 25

Fig. 2-2 Finite element model of the bridge in ANSYS ⋯⋯⋯⋯⋯⋯⋯⋯⋯ 25

Fig. 2-3 Model for three-axles vehicle ⋯⋯⋯⋯⋯⋯⋯⋯⋯⋯⋯⋯⋯⋯⋯ 26

Fig. 2-4 Deterioration of road roughness condition in a 15-year period ⋯⋯⋯⋯⋯ 28

Fig. 2-5 Cycles per truck passage ⋯⋯⋯⋯⋯⋯⋯⋯⋯⋯⋯⋯⋯⋯⋯⋯⋯ 33

Fig. 2-6 Fatigue reliability index for given vehicle speed and road roughness condition ⋯⋯ 37

Fig. 2-7 Fatigue life for given vehicle speed and road roughness condition ⋯⋯⋯⋯ 39

Fig. 3-1 Schematic of bridge responses (adapted from Billing, 1984) ⋯⋯⋯⋯⋯ 46

Fig. 3-2 Calculated dynamic responses of a three-span prestressed concrete bridge to a
four-axle vehicle (adapted from Green, 1993) ⋯⋯⋯⋯⋯⋯⋯⋯⋯⋯⋯ 46

Fig. 3-3 Typical section of bridge (unit= meter) ⋯⋯⋯⋯⋯⋯⋯⋯⋯⋯⋯⋯ 51

Fig. 3-4 Vehicle model for two axles ⋯⋯⋯⋯⋯⋯⋯⋯⋯⋯⋯⋯⋯⋯⋯⋯ 52

Fig. 3-5 Vehicle model for three axles ⋯⋯⋯⋯⋯⋯⋯⋯⋯⋯⋯⋯⋯⋯⋯ 53

Fig. 3-6 Vehicle model for five axles ⋯⋯⋯⋯⋯⋯⋯⋯⋯⋯⋯⋯⋯⋯⋯⋯ 55

Fig. 3-7 DAFS values for two axle truck ⋯⋯⋯⋯⋯⋯⋯⋯⋯⋯⋯⋯⋯⋯⋯ 59

Fig. 3-8 DAFS values for three axle truck ⋯⋯⋯⋯⋯⋯⋯⋯⋯⋯⋯⋯⋯⋯ 60

Fig. 3-9 DAFS values for five axle truck ⋯⋯⋯⋯⋯⋯⋯⋯⋯⋯⋯⋯⋯⋯ 61

Fig. 3-10 Parametric study of DAFC in life cycle ⋯⋯⋯⋯⋯⋯⋯⋯⋯⋯⋯⋯ 62

Fig. 3-11 Fatigue life estimation ⋯⋯⋯⋯⋯⋯⋯⋯⋯⋯⋯⋯⋯⋯⋯⋯⋯ 67

Fig. 3-12 Fatigue life versus DAFS ⋯⋯⋯⋯⋯⋯⋯⋯⋯⋯⋯⋯⋯⋯⋯⋯ 68

Fig. 4-1 Flowchart of the approach ⋯⋯⋯⋯⋯⋯⋯⋯⋯⋯⋯⋯⋯⋯⋯⋯ 78

Fig. 4-2 Typical section of bridge (unit= meter) ⋯⋯⋯⋯⋯⋯⋯⋯⋯⋯⋯⋯ 79

Fig. 4-3 Vehicle model for two axles·· 80

Fig. 4-4 Vehicle model for three axles·· 81

Fig. 4-5 Vehicle model for five axles ·· 82

Fig. 4-6 Equivalent stress ranges for 2 axle trucks ··· 83

Fig. 4-7 Equivalent stress ranges for 3 axle trucks ··· 84

Fig. 4-8 Equivalent stress ranges for 5 axle trucks ··· 84

Fig. 4-9 Comparison of fatigue damage evolution ·· 86

Fig. 4-10 Comparison of cumulative probability of failure ··· 86

Fig. 4-11 Fatigue life estimation for SD cases··· 87

Fig. 4-12 Fatigue damage evolution for SD cases ·· 88

Fig. 4-13 Cumulative probability of failure for SD cases ·· 89

Fig. 4-14 Fatigue lives for varied speed limit ·· 90

Fig. 4-15 Fatigue lives for varied COV ·· 91

Fig. 5-1 Bridge deck system of Donghai Bridge ··· 98

Fig. 5-2 Degrees of freedom of quadrilateral shell element··· 99

Fig. 5-3 Bridge deck section of the short-span bridge ·· 105

Fig. 5-4 FE model of the short-span bridge ·· 106

Fig. 5-5 Vehicle model for three axles·· 107

Fig. 5-6 Comparisons of static displacement of short-span bridge ··· 109

Fig. 5-7 Comparisons of static strain for short-span bridge ··· 110

Fig. 5-8 Comparisons of static stress for short-span bridge ··· 111

Fig. 5-9 Comparisons of dynamic displacemet history (100modes) for short-span bridge ····· 112

Fig. 5-10 Comparisons of dynamic stress history (100 modes) for short-span bridge ··········· 114

Fig. 5-11 Comparisons of dynamic stress history (19 modes) for short-span bridge ·············· 115

Fig. 6-1 Prototype bridge (Wu et al. 2003)··· 121

Fig. 6-2 Percentage of occurrence (adapted from Wu 2010) ·· 132

Fig. 6-3 Wind roses ··· 133

ABSTRACT

During the life cycle of bridges, varied amplitude of stress ranges on structural details are induced by the random traffic and wind loads. The progressive deteriorated road surface conditions might accelerate the fatigue damage accumulations. Micro-cracks in structural details might be initiated. An effective structural modeling scheme and a reasonable fatigue damage accumulation rule are essential for stress range acquisitions and fatigue life estimation. The present research targets at the development of a fatigue life and reliability prediction methodology for existing steel bridges under real wind and traffic environment with the capability of including multiple random parameters and variables in bridges' life cycle.

Firstly, the dynamic system is further investigated to acquire more accuracy stresses for fatigue life estimations for short and long span bridges. For short span bridges, the random effects of vehicle speed and road roughness condition are included in the limit state function, and fatigue reliability of the structural details is attained. For long-span bridges, a multiple scale modeling and simulation scheme based on the EOMM method is presented to obtain the stress range history of structural details, while the calculation cost and accuracy are saved. Secondly, a progressive fatigue reliability assessment approach based on a nonlinear continuous fatigue damage model is presented. At each block of stress cycles, types and numbers of passing vehicles are recorded to calculate the road surface's progressive deterioration and nonlinear cumulative fatigue damage, and the random road profiles are generated. Thirdly, this study discussed the fatigue design of short and long span bridges based on the dynamic analysis on the vehicle-bridge or vehicle-bridge-wind system. For short span bridges, a reliability-based dynamic amplification factor on revised equivalent stress ranges (DAFS) is proposed. For long span bridges, a comprehensive framework for fatigue reliability analysis under combined dynamic loads from vehicles and winds is presented. The superposed dynamic stress ranges cannot be ignored for fatigue reliability assessment of long-span bridges, although the stresses from either the vehicle loads or wind loads may not be able to induce serious fatigue issues alone.

8

CHAPTER 1   INTRODUCTION

This book consists of seven chapters. Chapter 1 introduces the related background knowledge of the dissertation, the research scope and structure of the dissertation. Chapter 2 discusses the random effects of vehicle speed and road roughness condition (Zhang and Cai 2011). Chapter 3 proposes a reliability-based dynamic amplification factor on stress ranges (DAFS) for fatigue design (Zhang and Cai 2012). Chapter 4 investigates bridge's progressive fatigue reliabilities based on a nonlinear continuous fatigue damage model. Chapter 5 discusses an equivalent orthotropic material modeling (EOMM) method. Chapter 6 investigates the combined dynamic loads effects on long-span bridges from winds and vehicles. Finally, Chapter 7 summarizes the dissertation and gives some suggestions for possible future research. More detailed descriptions of each chapter are given at the end of this chapter.

This introductory chapter gives a general background related to the present research. More detailed information can be found in each individual chapter.

1.1 Bridge Aerodynamics

Wind is the flow of air movement caused by differences in pressure. When structures immerse in the wind, the interactions between the wind and the structures might change the pressure distribution of structural surface or induce the structures to vibrate in a single or multiple frequencies. In addition, the location of bridges might be exposed to strong winds from wind storms, for instance, tropical cyclones (named as hurricanes in North America or typhoons in Asia-Pacific), thunderstorm, tornados, and downbursts (Chen 2004). Compared with earthquakes, wind loading produces roughly equal amounts of damage over a long period (Holmes 2001)

With the development of modern materials and construction techniques, the span length of bridges has reached to thousands of meters, such as suspension and cable-stayed bridges. Structural engineers and researchers have conducted various scientific investigations on bridge aerodynamics (Davenport 1962, Scanlan and Tomko 1971, Simiu and Scanlan 1996, Bucher and Lin 1988). Three approaches are currently used in the investigation of bridge aerodynamics: the wind tunnel experiment approach, the analytical approach, and the computational fluid dynamics approach (Chen 2004). As the backbones of the transportation lines in coastal areas and being vulnerable to wind loads, long-span bridges must be designed to withstand the drag forces induced by the mean wind, maintain dynamic stability under extreme wind conditions, and avoid serious fatigue failure under large wind induced vibrations due to aeroelastic effects. In order to investigate wind effects on structures analytically, wind induced vibrations were categorized as buffeting, flutter, galloping and vortex induced vibrations. The wind forces on bridges could be stated as the summation of static, self-excited, and buffeting force. Buffeting and vortex forces are similar, while the former is random vibrations and the latter is periodic vibrations. Under the dynamic effects from these two kinds of wind induced vibrations, fatigue damage would accumulate and may lead to an eventual collapse of bridges.

Before performing a time and space domain analysis of wind induced structural response, it is essential to generate the stochastic wind velocity field for numerical simulations. Many methods were proposed including the auto-regressive and moving-average (ARMA) models (Samaras et. al 1985, Naganuma et al 1987) and the spectral representation method. The latter can be extended to accommodate multidimensional non-Gaussian process cases (Yamazaki and Shinozuka, 1988). Based on the extension of the spectral representation method and the fast Fourier transform technique, Deodatis (1996) proposed an efficient computational method to generate ergodic sample functions of a stationary, multivariate stochastic process according to its prescribed cross-spectral density matrix. Cao et al. (2000) improved the algorithm and proposed the fast spectral representation method. It is improved by explicitly expressing Cholesky's decomposition of the cross-spectral density matrix in the form of algebraic formulas, then cutting off as many as possible of the cosine terms, so long as the accuracy of results is not affected. In addition, the fast Fourier transform technique was used to enhance the computational efficiency, as well. Chen and Letchford (2005) proposed a more effective method to simulate multivariate stationary Gaussian stochastic process by using a hybrid spectral representation and POD approach with negligible loss of accuracy. Later, Chen (2005) introduced the time-varying autoregressive (TVAR) model to develop a nonparametric deterministic-stochastic hybrid (NDESH) model to characterize and simulate non-stationary wind fields. Based on these methods, stochastic wind velocity histories can be generated for the applying wind forces, for instance, buffeting forces and self-excited forces, on the structure's finite element model.

Buffeting is defined as the forced random vibrations of a structure to random wind with a limited displacement and can only take place in turbulent flows. Buffeting response is random in nature and does not generally lead to structural failure but may cause serviceability or fatigue problems. Different with buffering forces, the self-excited forces induced by wind-structure interactions can cause flutter, which is one of the most dangerous aeroelastic phenomena for airfoils and large span bridges. Research on bridge buffeting analysis was initiated by Davenport (1961). In the 1960's, Davenport proposed the buffeting analysis method in frequency domain for large span bridges by introducing the statistical concepts of stationary time series and stochastic vibration theory. In Davenport's theory, the quasi-static linear theory was assumed to establish the buffeting forces and the aeroelastic damping. The aerodynamic admittance was used to take into account the effects of unsteadiness and spatial variation of wind turbulence surrounding the decks. In the 1970's, Scanlan and his co-workers proposed their buffeting response analysis method based on time-invariant linear system and aerodynamic strip theories (Scanlan and Gade 1977; Scanlan 1978). By simultaneously considering both the self-excited forces due to bridge deck motion and quasi-static linear aerodynamic forces caused by wind turbulence, the effects of both the aeroelastic stiffness and aeroelastic damping, and the aeroelastic coupling among translational and torsional vibrations can be taken into account in buffeting analysis. The buffeting forces are assumed to be linearly related to the fluctuations of wind speed, i.e., the aerodynamic coefficients are independent of wind turbulence.

Self-excited forces can be expressed in terms of some frequency-dependent flutter derivatives or time-dependent aerodynamic indicial functions. Scanlan used time-dependent aerodynamic indicial functions to express the self-excited forces that could be used in the time domain (Scanlan et al., 1974; Scanlan, 1984). The indicial functions can be either measured through wind tunnel tests or derived from the flutter derivatives obtained through section model wind tunnel test or numerical simulations. Since both of the flutter derivatives and wind spectra are the functions of frequency, the self-excited forces are usually expressed with the frequency-dependent flutter derivatives (Scanlan 1978; Scanlan and Jones 1990). In the time-domain analysis, the frequency dependent variables are difficult to be incorporated and frequency at time should be determined in order to quantify the self-excited force terms. Based on the linear assumption of self-excited forces, Lin and Yang (1983) expressed the self-excited forces in terms of convolution integrals between the bridge deck motion and the impulse response functions. The relationship between the aerodynamic impulse functions and flutter derivatives can be obtained by taking the Fourier transform of the expressions for the self-excited forces (Chen et al. 2000). As an alternative to the rational function approximation, complex eigenvalue analysis can also predict the vibration frequency iteratively for the dominant motion at any time step under any wind velocity (Chen and Cai 2003). The complex eigenvalue analysis can be conducted first to give the vibration frequency corresponding to each time through interactive process. The results can be incorporated into the coupled equations of motions to decide the self-excited force terms of the bridge.

In order to obtain the dynamic stress of bridge details, it is necessary to consider the spatial distribution of aerodynamic forces on bridge decks. In the work of Xu et al. (2009), the buffeting forces acting at the center of elasticity of the bridge deck are distributed to the nodes of the deck section in terms of the wind pressure distribution. In addition, the self-excited forces at the center of elasticity of the bridge deck can also be distributed to the nodes of the bridge deck by applying the virtual work principle (Chen 2010). In addition, surface wind pressure distributions can be measured from the wind tunnel experiments or via numerical simulations, as well.

1.2 Vehicle-Bridge-Wind Interaction Dynamics

Interaction analysis between vehicles and continuum structures originated in the middle of the 20th century. Initially, the vehicle loads were modeled as a constant moving force (Timoshenko et al. 1974) or a moving mass (Blejwas et al. 1979). The latter was used to consider inertial force. However, all the two models cannot include the effects of uneven bridge surface, which is known to be the main cause of high-magnitude bridge vibrations. Guo and Xu (2001) proposed a fully computerized approach for assembling equations of motions of coupled vehicle-bridge systems. In the dynamic system, vehicles are idealized as a combination of a number of rigid bodies connected by a series of springs and dampers. Later on, a fully computerized approach to simulate the interaction of the coupled vehicle-bridge system including a 3-D suspension vehicle model and a 3-D dynamic bridge model was developed (Shi et al. 2008). Direct integration method is used to treat the interaction by updating the characteristic matrices according to the position of contact points at each time step. Therefore, the equations of motion are time dependent and they should be modified, updated, and solved numerically by such as Runge-Kutta method at each time step. As an input to the updated matrix for the coupled equations of motions, the road surface roughness can be taken into account.

In the current AASHTO LRFD specifications (2010), the dynamic effects due to moving vehicles are attributed to two sources, namely, the hammering effect due to vehicle riding surface discontinuities, such as deck joints, cracks, potholes and delaminations, and dynamic response due to long undulations in the roadway pavement. A load roughness condition is usually quantified using Present Serviceability Rating (PSR), Road Roughness Coefficient (RRC) or International Roughness Index (IRI). Based on the studies carried out by Dodds and Robson (1973) and Honda et al. (1982), the long undulations in the roadway pavement could be assumed as a zero-mean stationary Gaussian random process and it could be generated through an inverse Fourier transformation (Wang and Huang 1992). For the surface discontinuities that cause hammer effects, these irregularities should be isolated and treated separately from such pseudo-random road surface profiles according to ISO (ISO 8606, 1995) and Cebon (1999). A twofold road surface condition can be used in the vehicle-bridge dynamic analysis to include the two sources for dynamic effects due to moving vehicles.

Based on studies on vehicle-bridge, vehicle-wind, and wind-bridge dynamics, Cai and Chen (2004) proposed a framework for the vehicle-bridge-wind aerodynamic analysis, which lay a very important foundation for vehicle accident analysis based on dynamic analysis results and facilitate the aerodynamic analysis of bridges considering vehicle-bridge-wind interaction. The framework built a general dynamic-mechanical model for vehicle-bridge-wind coupled system including both the structural part and loading part. The bridge and a series of vehicles can be simulated including various types of vehicles, while the external loading, like the wind effect and road roughness induced loading, can be included, as well. Based on full interaction analyses of a single-vehicle-bridge-wind system, the equivalent dynamic wheel load approach is proposed toward the study of fatigue performance of long-span bridges under both busy traffic and wind (Chen and Cai, 2007). Based on the detailed information of individual vehicles of stochastic traffic flow, Wu (2010) evaluated the lifetime performance of long-span bridges through taking account of more realistic traffic and wind environment effects. Given that many long-span bridges carry both trains and road vehicles, Chen (2010) carried the dynamic analysis of a coupled wind-train-road vehicle-bridge system.

With the calculated displacements, it is possible to analyze the stress responses and predict fatigue damages under combined dynamic loads from vehicles and winds at bridge details. Traditionally, a global structural analysis using a beam element model is first conducted to determine the critical locations. Based on the St. Venant's principle, the localized effects from loads will dissipate or smooth out with regions that are sufficiently away from the location of the load (Mises 1945). The forces are obtained from the beam element model and implemented only on a portion of the overall geometry to obtain the local static effects (Wu et al. 2003). The long-span bridges in kilometers long are usually built using beam elements and the model is usually called as a "fish-bone" type (Chan et al. 2008). However, only the rigid body motion is considered in the plane of the bridge deck section and the local deformations are neglected. After introducing the mixed dimensional coupling constraint equations developed by Monaghan (2000), the multi-scale model of Tsing Ma Bridge was built. Chan et al. (2005) merged a typical detailed joint geometry model into the beam element model to obtain the hot-spot stress concentration factors (SCF) of typical welded joints of the bridge deck. Then the hot spot stress block cycles were calculated by multiplying the nominal stress block cycles by the SCF for fatigue assessment. In order to model the bridges with multiple separated deck sections, such as the twin-box deck sections of the Stonecutters cable-stayed bridge and Xihoumen suspension bridge in China, two or more parallel "fish-spines" are suggested for the beam element model to model the bridge deck with multiple centroids of separate decks in order to obtain a reasonable result (Du 2006). Li et al. (2007) proposed a multi-scale FE modeling strategy for long-span bridges. The global structural analysis was carried out using the beam element modeling method at the level of a meter. The local detailed hot-spot stress analysis was carried out using shell or solid elements at the level of a millimeter. However, due to the limitations of the beam element modeling, the effects from distortion, constrained torsion, and shear lag were missing in the previous analyses, which might have a large effect on the local displacements, strains, and stresses for wide bridge decks with weak lateral connections. Nevertheless, in order to enhance the bending resistance of the steel plate to carry local loads from vehicle wheels, steel plates of the bridge decks are often stiffened with multiple closed or open stiffeners. Large computation efforts are needed for the refined section model with complicated structural details and it is difficult to include the time-varying dynamic effects from both wind loads and vehicle loads.

1.3 Bridge Fatigue

Fatigue is one of the main forms of deterioration for structures and can be a typical failure mode due to an accumulation of damage. During the life cycle of a bridge, the variable amplitude dynamic loading from vehicles on the deteriorated road surfaces can lead to fatigue damage accumulation in structure details. Such damages might develop into micro cracks and lead to serious fatigue failures for bridge components or a whole structure failure, for instance, the collapse and failure of the Point Pleasant Bridge in West Virginia (1967) and Yellow Mill Pond Bridge in Connecticut (1976).

Most structural components in steel bridges are assumed to be initially un-cracked. The stresses generated by repeated dynamic loadings are usually below the elastic limit of the structural steel. Therefore, the stress-based approach is widely used for fatigue analysis of steel bridges. If the stresses have constant amplitudes, the relations between the fatigue life and stress level can be achieved via coupon testing and S-N curves are obtained from the tests. In the current AASHTO

LRFD (2010) specifications, the S-N curve approach is adopted. Based on fatigue tests, Fisher et al. (1970) concluded that stress range and the type of weld details are the primary factors that influence the fatigue strength of steel bridge details. A large number of fatigue tests are carried out to construct the S-N curve. Based on the variability in the fatigue data, the S-N curve is defined as the 95% confidence limit for 95% survival of all details defined in each category. The fatigue life for a bridge detail can be obtained for the eight different categories. However, due to the overestimation of the stress ranges and the conservative manner in which the design curves for each category were defined, the design approach is conservative (Chung 2004). In addition, the numbers of cycles per truck passage were defined in a rather simply way and might underestimate the cycles under poor road roughness condition and high vehicle speed (Zhang and Cai 2011)

Nevertheless, the stresses generated by repeated dynamic loadings usually have variable amplitude ranges for most bridge details in practice. Compared with the fatigue under constant amplitude loadings, it is more difficult to model the fatigue problems correctly under variable amplitude loadings. Miner (1945) proposed the linear damage rule (LDR). Based on LDR, the equivalent stress range for variable amplitude ranges is obtained and S-N curve approach can be used for variable amplitude stress ranges. Even though LDR is most widely used for its simplicity, its shortcomings cannot be neglected. It may not be sufficient to describe the physics of fatigue damage accumulations (Fatemi and Yang 1998) and a large scatter in the fatigue life prediction can be found (Shimokawa and Tanaka 1980, Kawai and Hachinohe 2002, Yao et al 1986). During most of the length of bridges' fatigue lives, the structure materials are in a linear range and micro cracks have not developed into macroscopic cracks. After the initial crack propagation stage, the fatigue damage accumulation can be predicted through fracture mechanics analyses. However, the fatigue life assessment of existing bridges is related to a sequence of progressive fatigue damage with only the initiations of micro cracks. Nonlinear cumulative fatigue damage theories were developed to model the fatigue damage accumulation in this stage (Arnold and Kruch 1994, Chabache and Lesne 1988a, b). These theories are based either on separation of fatigue life into two periods (initiation and propagations) or on remaining life and continuous damage concepts. The nonlinear continuous fatigue damage model is more appropriate for the fatigue analysis during a large fraction of bridges' life cycle.

Considering the randomness inherent in both the load and resistance, reliability methods are appropriate for predicting structure lives based on the accumulated fatigue damages. Tang and Yao (1972) proposed a simple approach based on Miner's linear damage rule and it can treat the number of cycles leading to fatigue under various stress levels as a random variable. Later, Yao (1974) applied the fatigue reliability approach to the design of structural members with a specified acceptable probability of fatigue failure. Wirsching (1980) proposed a fatigue reliability analysis method for offshore structures to predict the fatigue failure at the welded joints under random wave loadings. Fatigue damage index D_f is first introduced in the S-N curve-based reliability analysis, which is now commonly used in fatigue reliability studies. Based on the Miner's linear damage rule, fatigue failure is defined as $D(t) > 1$ and the limit state function (LSF) is defined as (Nyman and Moses 1985):

$$g(X) = D_f - D(t) \qquad (1\text{-}1)$$

where D_f is the damage to cause failure and is treated as a random variable with a mean value of 1; $D(t)$ is the accumulated damage at time t and can be calculated based on the frequency domain analysis methods; and g is a failure function such that $g<0$ implies a fatigue failure. Simulation techniques can be used to solve the reliability problems. Monte Carlo method can be used to generate several results numerically without actually doing any physical testing. Since infinite simulations are impossible, a limited number of tests have to be accepted and the probability of failure can be obtained.

13

1.4 Overview of the Book

The main objective of this book is to develop a fatigue life and reliability prediction methodology for existing steel bridges under real wind and traffic environment for small, median and long-span bridges. A brief summary of each chapter of this book is provided next.

Chapter 2 discusses the random effects of vehicle speed and road roughness condition. Since each truck passage might generate multiple stress ranges, revised equivalent stress range is introduced to include fatigue damage accumulations for one truck passage. Therefore, the two variables, i.e., the stress range numbers and equivalent stress ranges per truck passage are coalesced in the new defined variable on a basis of equivalent fatigue damage. The revised equivalent stress range is obtained through a fully computerized approach toward solving a coupled vehicle-bridge system including a 3-D suspension vehicle model and a 3-D dynamic bridge model. At each truck-pass-bridge analysis, deteriorations of the road roughness condition are considered and the vehicle speed and road surface profile are generated randomly. Lognormal distribution is proven a good model to describe the revised equivalent stress range. In addition to the assumptions of other input random variables, fatigue reliability index and fatigue life for a target fatigue reliability index are predicted. The effects of the road surface condition, vehicle speed, and annual traffic increase rate on the fatigue reliability index and fatigue life are discussed, as well.

Chapter 3 proposes a reliability based dynamic amplification factor on stress ranges (DAFS) for fatigue design. A dynamic amplification factor (DAF) or dynamic load allowance (IM) is typically used in bridge design specifications to include dynamic effects from vehicles on bridges. The calculated live load stress ranges might not be correct due to varied dynamic amplification effects in different regions along the bridge, different road roughness conditions, and multiple stress range cycles generated for one vehicle passage on the bridge. Based on the revised equivalent stress defined in chapter 2, the fatigue damages from multiple stress ranges with varied amplitudes are equivalent to the fatigue damage from one stress cycle of the revised equivalent stress range. DAFS is then defined as the ratio of the nominal live load stress range and the maximum static stress range. A parametric study on DAFS is carried out to analyze the effect from multiple variables in the bridge's life cycle, for instance, faulting days in each year, vehicle speed limit and its coefficient of variance, vehicle type distribution, and annual traffic increase. In order to appreciate the difference of the proposed DAFS and traditional DAF, the calculated fatigue lives from the six approaches related to DAFS or DAF are compared with each other. Similar to DAF for dynamic response on displacements, DAFS is proposed to obtain dynamic stress ranges for fatigue design. As a result, once the DAFS is available, the dynamic stress ranges for fatigue design can be easily obtained by multiplying the maximum static stress range and the DAFS, which helps preserve both the accuracy and simplicity for bridge fatigue design.

Chapter 4 investigates bridge's progressive fatigue reliabilities based on a nonlinear continuous fatigue damage model. During most of the length of bridges' fatigue lives, the structure materials are in a linear range and micro cracks have not developed into macroscopic cracks. The fatigue life assessment of existing bridges is related to a sequence of progressive fatigue damage with only the initiations of micro cracks. Nonlinear cumulative fatigue damage theories were used to model the fatigue damage accumulation in this stage. It is more appropriate to use the nonlinear continuous fatigue damage model for the fatigue analysis during a large fraction of bridges' life cycle. Nevertheless, the road roughness conditions deteriorated with each repeated block of stress cycles induced by multiple vehicle passages and the vehicle types, numbers, and distributions might change with time, as well. Therefore, multiple random variables in the vehicle-bridge dynamic system during the bridge's life cycle are included in the proposed approach. Types and numbers of passing vehicles are recorded to calculate the road surface's progressive deterioration and road roughness coefficients are calculated for the each block of stress cycles. Fatigue damage accumulations and the cumulative probability of failures are calculated and recorded for each block of stress cycles. Once the threshold of road roughness coefficients is reached, the road profile is generated to the next category of the deteriorated road surface conditions or a road surface renovation is expected. The fatigue lives and fatigue damage index are obtained and compared with that obtained from linear fatigue damage model, as well.

Chapter 5 discusses an equivalent orthotropic material modeling (EOMM) method. Bridge details with complicated multiple stiffeners are modeled as equivalent shell elements using equivalent orthotropic materials, resulting in the same longitudinal and lateral stiffness in the unit width and shear stiffness in the shell plane as the original configuration. The static and dynamic response and dynamic properties of a simplified short span bridge from the EOMM model are obtained. The results match well with those obtained from the original model with real geometries and materials. The EOMM model for a long-span cable-stayed bridge is built with good precision on dynamic properties, which can be used for the wind induced fatigue analysis. Based on the modeling scheme, it is possible to predict a reasonable static and dynamic response of the bridge details due to the multi-scale dynamic loads effects, for instance, the wind induced vibrations of low frequency in kilo-meter scales and the vehicle induced vibrations of high frequency in meter scales.

Chapter 6 investigates combined dynamic loads effects on long-span bridges from winds and vehicles. After modeling the complicated structure details with equivalent orthotropic materials, dynamic stress ranges of a long-span bridge are obtained via solving the equations of motions for the vehicle-bridge-wind dynamic system with multiple random variables considered, for instance, vehicle speeds, road roughness conditions, and wind velocities and directions. After counting the number of stress cycles at different stress range levels using rainflow counting method, fatigue damage increments are obtained using the fatigue damage accumulation rule. The probability of failures for the fatigue damage at the end of each block of stress cycles and the cumulative probability of failures can be obtained. As a result, the fatigue life and reliability for the given structure details can be obtained. Based on the results from a case study, the dynamic effects from vehicles are found relatively small for long-span bridges and the effects from vehicle speeds and road roughness conditions can be neglected. Nevertheless, even though the stresses from either the vehicle loads or wind loads may not be able to induce serious fatigue problems alone, the superposed dynamic stress ranges cannot be ignored for the fatigue reliability assessment of long-span bridges.

Finally, Chapter 7 summarized the dissertation. Possible future research is recommended based on the current research.

1.5 References

AASHTO. (2010). "AASHTO LRFD bridge design specifications." Washington D.C.

Arnold, S. M., and Kruch, S. (1994). "A Differential CDM Model for Fatigue of Unidirectional Metal Matrix Composites." *International Journal of Damage Mechanics*, 3(2), 170-191.

Blejwas, T. E., Feng, C. C., and Ayre, R. S. (1979). "Dynamic interaction of moving vehicles and structures." *Journal of Sound and Vibration*, 67, 513-521.

Bucher, C. G., and Lin, Y. K. (1998). "Effects of spanwise correlation of turbulence field on the motion stability of long-span bridges." *Journal of Fluids and Structures*, 2(5), 437-451.

Cai, C. S., and Chen, S. R. (2004). "Framework of vehicle-bridge-wind dynamic analysis." *Journal of Wind Engineering and Industrial Aerodynamics*, 92(7-8), 579-607.

Cao, Y. H., Xiang, H. F., and Zhou, Y. (2000). "Simulation of stochastic wind velocity field on long-span bridges." *Journal of Engineering Mechanics*, 126(1), 1-6.

Cebon, D. (1999). Handbook of Vehicle-Road Interaction, Taylor & Francis.

Chaboche, J. L. (1988a). "Continuum Damage Mechanics: Part I---General Concepts." *Journal of Applied Mechanics*, 55(1), 59-64.

Chaboche, J. L. (1988b). "Continuum Damage Mechanics: Part II---Damage Growth, Crack Initiation, and Crack Growth." *Journal of Applied Mechanics*, 55(1), 65-72.

Chan, T. H. T., Zhou, T. Q., Li, Z. X., and Guo, L. (2005). "Hot spot stress approach for Tsing Ma Bridge fatigue evaluation under traffic using finite elment method." *Structural Engineering and Mechanics*, 19(3), 261-79.

Chan, T. H. T., Yu, Y., Wong, K. Y., & Li, Z. X. (2008). Condition-assessment-based finite element modeling of long-span bridge by mixed dimensional coupling method. (J. Gao, J. Lee, J. Ni, L. Ma, & J. Mathew, Eds.)Mechanics of Materials. Springer. Retrieved from http://eprints.qut.edu.au/16720/

Chen, L. (2005). "Vector time-varying autoregressive (TVAR) models and their application to downburst wind speeds," Ph.D Dissertation, Texas Tech University, Lubock.

Chen, L., and Letchford, C. W. (2005). "Simulation of multivariate stationary Gaussian stochastic processes: Hybrid spectral representation and proper orthogonal decomposition approach." *Journal of Engineering Mechanics*, 131(8), 801-808.

Chen, S. R. (2004). "Dynamic performance of bridges and vehicles under strong wind," Ph.D. dissertation, Louisiana State University, Baton Rouge.

Chen, S. R., and Cai, C. S. (2003). "Evolution of long-span bridge response to wind-numerical simulation and discussion." *Computers & Structures*, 81(21), 2055-2066.

Chen, S. R., and Cai, C. S. (2007). "Equivalent wheel load approach for slender cable-stayed bridge fatigue assessment under traffic and wind: Feasibility study." *Journal of Bridge Engineering*, 12(6), 755-764.

Chen, X., Matsumoto, M., and Kareem, A. (2000). "Time Domain Flutter and Buffeting Response Analysis of Bridges " *Journal of Engineering Mechanics*, 126(1).

Chen, Z. W. (2010). "Fatigue and Reliability Analysis of Multiload Suspension Bridges with WASHMS," Ph.D. Dissertation, The Hong Kong Polytechnic University, Hong Kong.

Chung, H. (2004). "Fatigue reliability and optimal inspection strategies for steel bridges," Ph.D. Dissertation, The University of Texas at Austin.

Davenport, A. G. (1961). "The application of statistical concepts to the wind loading of structures." Proceedings of the Institution of Civil Engineers, Structures and Buildings, 19, 449-472.

Davenport, A. G. (1962). "Buffeting of a suspension bridge by storm winds." *Journal of structural Division*, ASCE, 88(3), 233-268.

Deodatis, G. (1996). "*Simulation of ergodic multivariate stochastic processes.*" Journal of Engineering Mechanics, 122(8), 778-787.

Dodds, C. J., and Robson, J. D. (1973). "The Description of Road Surface Roughness." *Journal of Sound and Vibration*, 31(2), 175-183.

Du, B. (2006). "Dynamic Characteristics of Suspension Bridges with Twin-Box Girder Considering Nonlinearity," Ph.D Dissertation, Tongji University, Shanghai.

Fatemi, A., and Yang, L. (1998). "Cumulative fatigue damage and life prediction theories: a survey of the state of the art for homogeneous materials." *International Journal of Fatigue*, 20(1), 9-34.

Fisher, J. W., Frank, K. H., Hirt, M. A., and McNamee, B. M. (1970). "Effect of Weldments on the Fatigue Strength of Steel Beams." National Cooperative Highway Research Program Report 102, Transportation Research Board, National Research Council, Washington, D.C.

Guo, W. H., and Xu, Y. L. (2001). "Fully computerized approach to study cable-stayed bridge-vehicle interaction." *Journal of Sound and Vibration*, 248(4), 745-761.

Holmoes, J.D. (2001). *Wind loading of structures*, Spon Press.

Honda, H., Kajikawa, Y., and Kobori, T. (1982). "Spectra Of Road Surface Roughness On Bridges." *Journal of the Structural Division*, 108(ST-9), 1956-1966.

ISO. (1995). "Mechanical vibration - Road surface profiles - Reporting of measured data." Geneva.

Kawai, M., and Hachinohe, A. (2002). "Two-stress level fatigue of unidirectional fiber-metal hybrid composite: GLARE 2." *International Journal of Fatigue*, 24, 567-80.

Li, Z. X., Zhou, T. Q., Chan, T. H. T., and Yu, Y. (2007). "Multi-scale numerical analysis on dynamic response and local damage in long-span bridges." *Engineering Structures*, 29(7), 1507-1524.

Lin, Y. K., and Yang, J. N. (1983). "Multimode Bridge Response to Wind Excitations." *Journal of Engineering Mechanics*, 109(2), 586-603.

Mises, R. v. (1945). "On Saint Venant's principle." *Bulletin of the American Mathematical Society*, 51(3), 555-562.

Monaghan, D. J. (2000). "Automatically coupling elements of dissimilar dimension in finite element analysis," Ph.D Thesis, The Queen's University of Belfast.

Naganuma, T., and Deodatis, G. (1987). "ARMA Model for Two-Dimensional Processes." *Journal of Engineering Mechanics*, 113, 234-251.

Nyman, W. E., and Moses, F. (1985). "Calibration of Bridge Fatigue Design Model." *Journal of Structural Engineering*, 111(6), 1251-1266.

Samaras, E., Shinzuka, M., and Tsurui, A. (1985). "ARMA Representation of Random Processes." *Journal of Engineering Mechanics*, 111(3), 449-461.

Scanlan, R. H. (1978). "The action of flexible bridges under wind I: Flutter theory; II: Buffeting theory." *Journal of Sound and Vibration*, 60(2), 187-199; 201-211.

Scanlan, R. H. (1984). "Role of Indicial Functions in Buffeting Analysis of Bridges." *Journal of Structural Engineering*, 110(7), 1433-1446.

Scanlan, R. H., Béliveau, J.-G., and Budlong, K. S. (1974). "Indicial Aerodynamic Functions for Bridge Decks." *Journal of the Engineering Mechanics Division*, 100(4), 657-672.

Scanlan, R. H., and Gade, R. H. (1977). "Motion of suspended bridge spans under gusty wind." *Journal of Structure Division*, ASCE, 103(ST9), 1867-1883.

Scanlan, R. H., and Jones, N. P. (1990). "Aeroelastic Analysis of Cable-Stayed Bridges." *Journal of Structural Engineering*, 116(2), 279-297.

Scanlan, R. H., and Tomko, J. J. (1971). "Airfoil and bridge flutter derivatives." *Journal of Engineering Mechanics Division*, ASCE, 97(6), 1717-1737.

Shi, X., Cai, C. S., and Chen, S. (2008). "Vehicle Induced Dynamic Behavior of Short-Span Slab Bridges Considering Effect of Approach Slab Condition." *Journal of Bridge Engineering*, 13(1), 83-92.

Shimokawa, T., and Tanaka, S. (1980). "A statistical consideration of Miner's rule." *International Journal of Fatigue*, 4, 165-70.

Simiu, E., and Scanlan, R. H. (1996). Wind effects on structures: an introduction to wind engineering, John Wiley & Sons, NewYork.

Tang, J. P., and Yao, J. T. P. (1972). "Fatigue Damage Factor in Structural Design." *Journal of the Structural Division*, 98(1), 125-134.

Timoshenko, S., Young, D. H., and Weaver, W. (1974). Vibration problems in engineering, Wiley, New York.

Wang, T.-L., and Huang, D. (1992). "Computer modeling analysis in bridge evaluation." Florida Department of Transportation, Tallahassee, FL.

Wirsching, P. H. (1980). "Fatigue reliability in welded joints of offshore structures." *International Journal of Fatigue*, 2(2), 77-83.

Wu, C., Zeng, M.G. and Dong, B. (2003) "Report on the Performance of Steel-Concrete Composite Beam of Donghai Cable-stayed Bridge." Department of Bridge Engineering, Tongji University, Shanghai.

Wu, J. (2010). "Reliability-Based Lifetime Performance Analysis of Long-span Bridges," Ph.D. Dissertation, Colorado State University, Fort Collins, Colorado.

Xu, Y. L., Liu, T. T., and Zhang, W. S. (2009). "Buffeting-induced fatigue damage assessment of a long suspension bridge." *International Journal of Fatigue*, 31(3), 575-586.

Yamazaki, F., Shinozuka, M., and Dasgupta, G. (1988). "Neumann Expansion for Stochastic Finite Element Analysis." *Journal of Engineering Mechanics*, 114, 1335-1354.

Yao, J. T. P. (1974). "Fatigue Reliability and Design." *Journal of the Structural Division*, 100(9), 1827-1836.

Yao, J. T. P., Kozin, F., Wen, Y. K., Yang, J. N., Schueller, G. I., and Ditlevsen, O. (1986). "Stochastic fatigue fracture and damage analysis." *Structural Safety*(3), 231-67.

Zhang, W., C.S. Cai. (2011). "Fatigue Reliability Assessment for Existing Bridges Considering Vehicle and Road Surface Conditions", *Journal of Bridge Engineering*, doi:10.1061/ (ASCE) BE. 1943-5592.0000272

Zhang, W., C.S. Cai. (2012). "Reliability Based Dynamic Amplification Factor on Stress Ranges for Fatigue Design of Existing Bridges", *Journal of Bridge Engineering*, doi:10.1061/ (ASCE) BE. 1943-5592.0000387

CHAPTER 2   FATIGUE RELIABILITY ASSESSMENT FOR EXISTING BRIDGES CONSIDERING VEHICLE SPEED AND ROAD SURFACE CONDITIONS

2.1 Introduction

After the interstate-35 Bridge in the state of Minnesota collapsed in August 2007, concerns about the safety and risk assessment of existing bridges have been greatly increased. During the life cycle of a bridge, dynamic impacts due to random traffic loads and deteriorated road surface conditions can induce serious fatigue issues for bridge components. It is necessary and realistic to use reliability method and treat the input parameters as random variables for the vehicle-bridge dynamic system. Decisions, such as structure replacement, deck replacement or some other retrofit measures, can be made based on estimated fatigue reliability index to ensure structure safety and normal service condition.

In fatigue design, the load-induced fatigue effect should be less than the nominal fatigue resistance. Naturally, the fatigue requirement can also be stated as the fatigue life consumed by the load being less than the available fatigue life of the bridge detail. However, in a typical fatigue analysis in current specifications such as American Association of State Highway and Transportation Officials (AASHTO) Load and Resistance Factor Design (LRFD) bridge design specifications (2007), vehicle speeds and road roughness conditions are not considered, which have been proven to have significant effects on the dynamic responses of short span bridges (Deng and Cai 2010; Shi et al. 2008).

In order to obtain stress range history, a data analysis on on-site strain measurements or a structural dynamic analysis of bridges is necessary. Since the field measurements can be expensive and stress range spectra for bridges are strongly site-specific (Laman and Nowak 1996), it is impossible to take on-site measurement for every concerned location of every bridge. Nevertheless, Finite Element Method (FEM) based structural dynamic analysis can provide reasonable stress range history for bridge details in various scenarios. During the analysis, the vehicles loads were modeled from a constant moving force (Timoshenko et al. 1974), moving mass (Blejwas et al. 1979) to through a full vehicle-bridge coupled model (Guo and Xu 2001). In the coupled model, the contact point changes all the time when the vehicle travels the bridge. After including the wind induced forces on bridge and vehicles, a vehicle-bridge-wind interaction model was proposed to study the dynamic performance of the coupled system and related vehicle accident risks of overturning and side slipping (Cai and Chen 2004). Later on, in order to consider the 3D effects for short and wide bridges, the vehicle bridge dynamic system was improved to include 3-D model for both the vehicles and the bridges (Shi et al. 2008). Both of the 3D models for vehicles and bridges make it possible to obtain stress history and carry fatigue reliability assessment for any designated bridge details.

It is noted that variations of structural properties and static loading are usually considered in a typical structural reliability analysis. In the present study, these factors are not discussed since they have been widely covered in the literature. The present study aims to demonstrate how to deal with the dynamic live load effects that are affected by the randomness of vehicle speeds and road surface profiles in a bridge's service life. Therefore, a framework of fatigue reliability assessment for existing bridges is proposed considering the random effects of vehicle speed and road roughness condition. The developed methodology can be used to assess the fatigue life of site specific bridges in the context of available reliability analyses. Assumptions are made to simplify the framework to make it manageable. In the present study, the lateral position of the vehicle, the deterioration of the road surface and the resurface scheme and the traffic increase rate are considered. In the future study, more parameters need to be considered as random variables for actual applications

2.2 Vehicle-Bridge Dynamic System

2.2.1. Vehicle and Bridge Model

In the present study, the vehicle is modeled as a combination of several rigid bodies connected by several axle mass blocks, springs and damping devices (Cai and Chen 2004). The tires and suspension systems are idealized as linear elastic spring elements and dashpots. The vehicles with axle number from two to five can be simulated using the model.

The equation of motion for the vehicle is derived based on the following matrix form:

$$[M_v]\{\ddot{d}_v\}+[C_v]\{\dot{d}_v\}+[K_v]\{d_v\}=\{F_v^G\}+\{F_c\} \qquad (2\text{-}2)$$

where, $[M_v]$, the mass matrix, $[C_v]$, damping matrix and $[K_v]$, stiffness matrix are obtained by considering the equilibrium of the forces and moments of the system; $\{F_v^G\}$ is the self-weight of vehicle; $\{F_c\}$ is the vector of wheel-road contact forces acting on the vehicle.

The dynamic model of bridges can be obtained through finite element method using different finite elements such as beam, solid or shell elements. The motion of the bridge can be stated as the following equations:

$$[M_b]\{\ddot{d}_b\}+[C_b]\{\dot{d}_b\}+[K_b]\{d_b\}=\{F_b\} \qquad (2\text{-}1)$$

where, $[M_b]$ is the mass matrix, $[C_b]$ is the damping matrix and $[K_b]$ is the stiffness matrix of the bridge, and $\{F_b\}$ is wheel-bridge contact forces on bridge.

The stiffness matrix might change with the fatigue damages and other deteriorative damages, which results in a change of the bridge frequencies and modal shapes (Salane and Baldwin, 1990). However, this change is expected to be small. In order to consider the effects of the road surface condition and vehicle speed on fatigue reliability of bridges, the bridge stiffness matrix were assumed as constants in the present study. The equations of motion for the vehicle and bridge are coupled through the interaction force, i.e. F_b and F_c. F_b and F_c are action and reaction forces existing at the contact points of the two systems.

2.2.2. Interactions between Vehicle and Bridge

The equations of motion for the vehicle and bridge are listed in Eqs. (2-1) and (2-2). However, in order to solve the equations, it is required to calculate the forces on the right side of the equations, namely, the contact forces between vehicle and bridge.

Based on previous work (Cai and Chen 2004), the interactions between the bridge and vehicles are modeled as coupling forces between the vehicle tires and the road surface. The contact forces can be stated as a function of deformation of the vehicle's lower spring:

$$\{F_b\}=-\{F_c\}=[K_l]\{\Delta_l\}+[C_l]\{\dot{\Delta}_l\} \qquad (2\text{-}2)$$

where, $[K_l]$ and $[C_l]$ are coefficients of vehicle lower spring and damper; and Δ_l is deformation of lower springs of vehicle. The relationship among vehicle-axle-suspension displacement Z_a, displacement of bridge at wheel-road contact points Z_b, deformation of lower springs of vehicle Δ_l, and road surface profile $r(x)$ is:

$$Z_a = Z_b + r(x) + \Delta_l \tag{2-3}$$

$$\dot{Z}_a = \dot{Z}_b + \dot{r}(x) + \dot{\Delta}_l \tag{2-4}$$

where $\dot{r}(x) = \left(dr(x)/dx\right)\cdot\left(dx/dt\right) = \left(dr(x)/dx\right)\cdot V(t)$ and $V(t)$ is the vehicle velocity.

Therefore, the contact force F_b and F_c between the vehicle and the bridge is:

$$\{F_b\} = -\{F_c\} = [K_l]\{Z_a - Z_b - r(x)\} + [C_l]\{\dot{Z}_a - \dot{Z}_b - \dot{r}(x)\} \tag{2-5}$$

2.2.3.   Mode Superposition Techniques

After transforming the contact forces to equivalent nodal force and substituting them into Eqs. (2-1) and (2-2), the final equations of motion for the coupled system are as follows (Shi et al. 2008):

$$\begin{bmatrix} M_b & \\ & M_v \end{bmatrix}\begin{Bmatrix} \ddot{d}_b \\ \ddot{d}_v \end{Bmatrix} + \begin{bmatrix} C_b + C_{bb} & C_{bv} \\ C_{vb} & C_v \end{bmatrix}\begin{Bmatrix} \dot{d}_b \\ \dot{d}_v \end{Bmatrix} + \begin{bmatrix} K_b + K_{bb} & K_{bv} \\ K_{vb} & K_v \end{bmatrix}\begin{Bmatrix} d_b \\ d_v \end{Bmatrix} = \begin{Bmatrix} F_{br} \\ F_{vr} + F_v^G \end{Bmatrix} \tag{2-6}$$

The additional terms C_{bb}, C_{bv}, C_{vb}, K_{bb}, K_{bv}, K_{vb}, F_{br} and F_{vr} in Eq. (2-7) are due to the expansion of the contact force in comparison with Eqs. (2-1) and (2-2). When the vehicle is moving across the bridge, the bridge-vehicle contact points change with the vehicle position and the road roughness at the contact point. As a large number of degrees of freedom (DOF) are involved, the bridge mode superposition technique is used to simplify the modeling procedure based on the obtained bridge mode shape $\{\Phi_i\}$ and the corresponding natural circular frequencies ω_i. Bridge fatigue analysis corresponds to service load level and the bridge performance is practically in the linear range, which justifies the use of the modal superposition approach. Consequently, the number of equations in Eq. (2-7) and the complexity of the whole procedure are greatly reduced.

If each mode shape is normalized to mass matrix, i.e. $\{\Phi_i\}^T[M_b]\{\Phi_i\}=1$ and $\{\Phi_i\}^T[K_b]\{\Phi_i\}=\omega_i^2$, and if the damping matrix $[C_b]$ is assumed to be $2\omega_i\eta_i[M_b]$, where ω_i is the natural circular frequency of the bridge and η_i is the percentage of the critical damping for the bridge i^{th} mode, Eq. (2-7) can be rewritten as (Shi et al. 2008):

$$\begin{bmatrix} I & \\ & M_v \end{bmatrix}\begin{Bmatrix} \ddot{\xi}_b \\ \ddot{d}_v \end{Bmatrix} + \begin{bmatrix} 2\omega_i\eta_i I + \Phi_b^T C_{bb}\Phi_b & \Phi_b^T C_{bv} \\ C_{vb}\Phi_b & C_v \end{bmatrix}\begin{Bmatrix} \dot{\xi}_b \\ \dot{d}_v \end{Bmatrix} + \begin{bmatrix} \omega_i^2 I + \Phi_b^T K_{bb}\Phi_b & \Phi_b^T K_{bv} \\ K_{vb}\Phi_b & K_v \end{bmatrix}\begin{Bmatrix} \xi_b \\ d_v \end{Bmatrix} = \begin{Bmatrix} \Phi_b^T F_{br} \\ F_{vr} + F_v^G \end{Bmatrix} \tag{2-7}$$

The bridge dynamic response $\{d_b\}$ can be expressed as:

$$\{d_b\} = [\Phi_b]\{\xi_b\} = \left[\{\Phi_1\} \quad \{\Phi_2\}...\{\Phi_n\}\right]\{\xi_1 \quad \xi_2 \cdots \xi_n\}^T \tag{2-8}$$

where n is the total number of modes for the bridge under consideration; $\{\Phi_i\}$ and ξ_i are the ith mode shape and its generalized coordinates. The stress vector can be obtained by:

$$[S] = [E][B]\{d_b\} \tag{2-9}$$

where, $[E]$ is the stress-strain matrix and is assumed to be constant over the element and $[B]$ is the strain-displacement matrix assembled with x, y and z derivatives of the element shape functions.

22

2.3 Modeling of Progressive Deterioration for Road Surface

2.3.1. Generation of Road Surface Roughness Spectra

Road surface roughness is generally defined as an expression of irregularities of the road surface and it is the primary factor affecting the dynamic response of both vehicles and bridges (Deng and Cai 2010; Shi et al. 2008). Based on the studies carried out by Dodds and Robson (1973) and Honda et al. (1982), the road surface roughness was assumed as a zero-mean stationary Gaussian random process and it could be generated through an inverse Fourier transformation as (Wang and Huang 1992):

$$r(x) = \sum_{k=1}^{N} \sqrt{2\phi(n_k)\Delta n} \cos(2\pi n_k x + \theta_k) \tag{2-10}$$

where θ_k is the random phase angle uniformly distributed from 0 to 2π; $\phi()$ is the power spectral density (PSD) function (m^3/cycle/m) for the road surface elevation; n_k is the wave number (cycle/m). The PSD functions for road surface roughness were developed by Dodds and Robson (1973) and three groups of road classes were defined with the values of roughness exponents ranging from 1.36 to 2.28 for motorways, principal roads and minor roads. In order to simplify the description of road surface roughness, both the two roughness exponents were assumed to have a value of two and the PSD function was simplified by Wang and Huang (1992) as:

$$\phi(n) = \phi(n_0)(\frac{n}{n_0})^{-2} \tag{2-11}$$

where $\phi(n)$ is the PSD function (m^3/cycle) for the road surface elevation; n is the spatial frequency (cycle/m); n_0 is the discontinuity frequency of $1/2\pi$ (cycle/m); and $\phi(n_0)$ is the road roughness coefficient (m^3/cycle) and its value is chosen depending on the road condition.

2.3.2. Road Roughness Index

Road roughness condition is classically quantified using Present Serviceability Rating (PSR), Road Roughness Coefficient (RRC) or International Roughness Index (IRI). Both of the Present Serviceability Rating (PSR) and Road Roughness Coefficient (RRC) classified the road roughness condition as very good, good, fair (average), poor and very poor. The PSR was based on passenger interpretations of ride quality, which is developed by the AASHTO Road Test. The subjective scale ranges were set from five (excellent) to zero (essentially impassable). The International Organization for Standardization (1995) used RRC to define the road roughness classification and the ranges were listed in Table 2-1. It is noteworthy that RRC is based on the road profiles only. The international roughness index (IRI) was developed in 1986 and is used to define the longitudinal profile of a traveled wheel track (Sayers and Karamihas 1998). The IRI is based on the average rectified slope (ARS), which is a filtered ratio of a standard vehicle's accumulated suspension motion divided by the distance traveled by the vehicle during the measurement. Various correlations have been developed between the indices (Paterson 1986; Shiyab 2007). Based on the corresponding ranges of the road roughness coefficient and the IRI value (Shiyab 2007), a relationship between the IRI and the RRC is utilized in the present study:

$$\phi(n_0) = 6.1972 \times 10^{-9} \times e^{IRI/0.42808} + 2 \times 10^{-6} \tag{2-12}$$

Table 2-1 RRC values for road roughness classifications

Road roughness classifications	Ranges for RRCs
Very good	2×10^{-6}- 8×10^{-6}
Good	8×10^{-6}- 32×10^{-6}
Average	32×10^{-6}-128×10^{-6}
Poor	128×10^{-6}- 512×10^{-6}
Very poor	512×10^{-6}- 2048×10^{-6}

2.3.3. Progressive Deterioration Model for Road Roughness

In order to consider the road surface damages due to loads or corrosions, a progressive deterioration model for road roughness is necessary. More specifically, it is essential to have such a model for RRC in order to generate the random road profile.

IRI values at any time after the service of road surface are calculated as (Paterson 1986):

$$IRI_t = 1.04e^{\eta \cdot t} \cdot IRI_0 + 263(1+SNC)^{-5}(CESAL)_t \qquad (2\text{-}13)$$

where IRI_t is the IRI value at time t; IRI_0 is the initial roughness value directly after completing the construction and before opening to traffic; t is the time in years; η is the environmental coefficient varying from 0.01 to 0.7 that depends on the dry or wet, freezing or non-freezing conditions; Structural number (SNC) is a parameter that is calculated from data on the strength and thickness of each layer in the pavement and $(CESAL)_t$ is the estimated number of traffic in terms of AASHTO 80-kN(18-kip) cumulative equivalent single axle load at time t in millions.

Therefore, the RRC at any time after construction is predicted using Eqs. (2-13) and (2-14):

$$\phi(n_0)_t = 6.1972\times10^{-9} \times \exp\left\{\left[1.04e^{\eta t} \cdot IRI_0 + 263(1+SNC)^{-5}(CESAL)_t\right]/0.42808\right\} + 2\times10^{-6} \qquad (2\text{-}14)$$

2.4 Prototypes of Bridge and Vehicle

2.4.1. Prototype of the Bridge

To demonstrate the methodology of fatigue reliability assessment of existing bridges due to vehicle-induced dynamic responses, a short span slab-on-girder bridge designed in accordance with AASHTO LRFD bridge design specifications (AASHTO 2007) is analyzed. The bridge has a span length of 12 m (39.4 ft) and a width of 13m (42.7ft), which accommodates two vehicle lanes traveling in the same direction. The concrete deck is 0.18m (7 in) thick and the haunch is 0.04m (1.6 in) high. All of the seven steel girders are W27×94 and have an even spacing of 2m (6.6 ft) as shown in Fig. 2-1. The intermediate and end cross-frames enable the girders to deflect more equally. In this bridge, a steel channel section, C15×33.9, is used as cross-frame. In the present study, after conducting a sensitivity studying by changing the meshing, 27543 solid elements and 43422 nodes are used to build the finite element model of the bridge. The whole model is shown in Fig.2-2. The damping ratio is assumed to be 0.02. The present study focuses on the fatigue analysis at the longitudinal welds located at the conjunction of the web and the bottom flange at the mid-span.

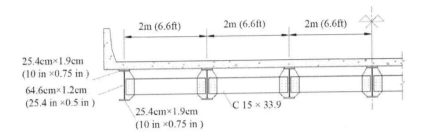

Fig. 2-1.　Typical section of bridge (unit= meter)

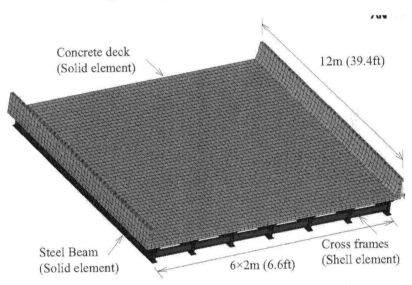

Fig. 2-2.　Finite element model of the bridge in ANSYS

2.4.2.　Prototype of the Vehicle

Many methods have been used to simulate the traffic flow to obtain the load effects for short to medium span bridges including white noise fields (Ditlevsen and Madsen 1994), Poisson's distribution (Cheng et al. 2006) and Monte Carlo approach (Moses 2001; O'Connor and O'Brien 2005). In order to obtain the actual truckload spectra, weigh-in-motion (WIM) techniques have been developed and extensively used nationwide. Based on the data from WIM measurements, fifteen vehicle types are defined according to Federal Highway Administration (FHWA) classification scheme "F". Since the design live load for the prototype of the bridge is HS20-44 truck, this three-axle truck is chosen as the prototype of the vehicle in the present study. Several vehicles that travel on the bridge may have different speeds and may be located in different lanes randomly or simultaneously. The common practice is to use only one vehicle or a series of identical vehicles in one lane (Guo and Xu 2001). In the present study, only one vehicle in one lane is considered to travel

25

along the bridge for fatigue analysis due to its short span length. Based on the strategy that was used in developing the AASHTO LRFD bridge design specifications (2007) for fatigue design, occasional presence of other trucks on the bridge will not significantly affect the fatigue life of bridges. A 6 m (39.4 ft) long approach slab connecting the pavement and bridge deck is considered. The three-axle truck model used in present study is shown in Fig. 2-3. The geometry, mass distribution, damping, and stiffness of the tires and suspension systems of the truck are listed in Table 2-2.

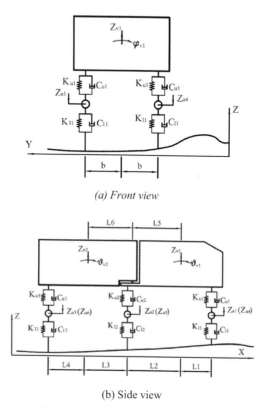

(a) Front view

(b) Side view

Fig. 2-3.   Model for three-axles vehicle

26

Table 2-2 Major parameters of vehicle (3 axles)

Mass	truck body 1	2612 kg (5746 lbs)
	truck body 2	26113 kg (57448 lbs)
	first axle suspension	490 kg (1078 lbs)
	second axle suspension	808 kg (1777 lbs)
	third axle suspension	653 kg (1436 lbs)
Moment of inertia	Pitching, truck body1	2022 kg.m2 (47882 lbs.ft2)
	Pitching, truck body2	33153 kg.m2 (785083 lbs.ft2)
	Rolling, truck body2	8544 kg.m2 (202327 lbs.ft2)
	Rolling, truck body2	181216 kg.m2 (4291304 lbs.ft2)
Spring stiffness	Upper, 1st axle	242604 N/m (16623 lbs/ft)
	Lower, 1st axle	875082 N/m (59962 lbs/ft)
	Upper, 2nd axle	1903172 N/m (130408 lbs/ft)
	Lower, 2nd axle	3503307 N/m (240052 lbs/ft)
	Upper, 3rd axle	1969034 N/m (134921 lbs/ft)
	Lower, 3rd axle	3507429 N/m (240335 lbs/ft)
Damping coefficient	Upper, 1st axle	2190 N.s/m (150 lbs.s/ft)
	Lower, 1st axle	2000 N.s/m (137 lbs.s/ft)
	Upper, 2nd axle	7882 N.s/m (540 lbs.s/ft)
	Lower, 2nd axle	2000 N.s/m (137 lbs.s/ft)
	Upper, 3rd axle	7182 N.s/m (492 lbs.s/ft)
	Lower, 3rd axle	2000 N.s/m (137 lbs.s/ft)
Length	L1	1.698 m (5.6 ft)
	L2	2.569 m (8.4 ft)
	L3	1.984 m (6.5 ft)
	L4	2.283 m (7.5 ft)
	L5	2.215 m (7.3 ft)
	L6	2.338 m (7.7 ft)
	B	1.1 m (3.6 ft)

2.4.3. Modeling of Vehicle Speed

The dynamic displacement of bridges was found changing with the vehicle speed based on previous studies (Cai and Chen 2004; Cai et al. 2007). Typically, the maximum speed limits posted in bridges or roads are based on the 85th percentile speed when adequate speed samples are available. The 85th percentile speed is a value that is used by many states and cities for establishing regulatory speed zones (Donnell et al. 2009; TxDOT 2006). Statistical techniques show that a normal distribution occurs when random samples of traffic are measured (TxDOT 2006). This allows describing the vehicle speed conveniently with two characteristics, i.e. the mean and standard deviation. In the present study, the 85th percentile speed is approximated as the sum of the mean value and one standard deviation for simplification. The speed limit is assumed as 26.8m/s (60mph) and the coefficient of variance of vehicle speeds is assumed as 0.4. In order to simplify the calculations, the randomly generated vehicle speeds are grouped into six ranges that are represented by the vehicle speed from 10m/s (22.4 mph) to 60m/s (134.4mph). The probabilities of vehicle speed in all ranges are listed in Table 2-3.

27

Table 2-3 Vehicle speed ranges

U_{ve}	Vehicle speed range	Probability
10m/s (22.4mph)	<15m/s (33.6 mph)	2.9575E-01
20m/s (44.8mph)	15m/s (33.6mph) - 25m/s (56mph)	4.8426E-01
30m/s (67.2mph)	25m/s (56mph) -35m/s(78.4 mph)	2.0127E-01
40m/s (89.6mph)	35m/s (78.4 mph) - 45m/s (100.8 mph)	1.8361E-02
50m/s (112mph)	45m/s (100.8 mph) - 55m/s (123.2 mph)	3.4809E-04
60m/s (134.4 mph)	>55m/s (123.2 mph)	1.3075E-06

2.4.4. Modeling of Road Roughness

For existing bridges, the past records of road roughness conditions can be tracked and the future conditions can be predicted based on the history records. In the present demonstration study, the equations for road roughness coefficients are formulated by assuming $SNC = 4$, $\eta = 0.1$. The total Average Daily Truck Traffics (ADTTs) for trucks are distributed to the two lanes of the bridge with the fast lane occupying 30% and the slow lane 70%. In the present study, the ADTT for the first year is assumed as 2000. If the traffic increase rate α is 0%, the $CESAL$ for the fast and slow lane for each year is 413,362 and 964,513, respectively. The RRCs and IRIs for the two lanes in 15 years after construction are shown in Fig. 2-4 (a) and (b). The two progressive deterioration functions of the road roughness coefficients for the two lanes are thus defined. After 15 years, a surface renovation is expected. In order to simplify the calculations, the road roughness condition is grouped into ranges from very good to very poor and a typical value of road roughness coefficient is chosen to represent the range.

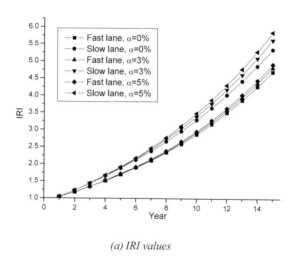

(a) IRI values

Fig. 2-4.   Deterioration of road roughness condition in a 15-year period

28

(Fig. 2-4 continued)

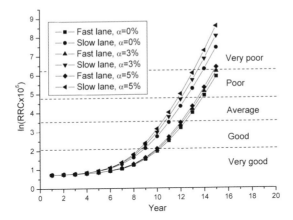

(b) RRC values

Based on the RRC calculated from Eq. (2-15), for the fast lane in a fifteen-year period, the road condition in the first ten years is classified as very good, the eleventh and twelfth years as good, the thirteenth year as average , and the fourteenth and fifteenth   years as poor:

$$\phi(n_0)_t = \begin{cases} 5\times10^{-6} & 1 \le t \le 10 \ years \\ 20\times10^{-6} & 11 \le t \le 12 \, years \\ 80\times10^{-6} & t = 13 \ years \\ 320\times10^{-6} & 14 \le t \le 15 \ years \end{cases}$$ (2-15)

Similarly, for the slow lane, the road condition is defined as very good in the first eight years, good in the ninth and tenth years, average in the eleventh and twelfth years, poor in the thirteen year, and very poor in the fourteen and fifteen years:

$$\phi(n_0)_t = \begin{cases} 5\times10^{-6} & 1 \le t \le 8 \ years \\ 20\times10^{-6} & 9 \le t \le 10 \, years \\ 80\times10^{-6} & 11 \le t \le 12 \ years \\ 320\times10^{-6} & t = 13 \ years \\ 1280\times10^{-6} & 14 \le t \le 15 \, years \end{cases}$$ (2-16)

Due to the traffic increase per year, the CESAL changes, and results in a change of the progressive deterioration function. Based on the ADTT and traffic increase rate per year, the cumulated number of truck passages for the future year t is estimated (Kwon and Frangopol, 2010):

$$n_{tr}(t) = 365 \cdot ADTT \cdot \int_0^t (1+\alpha)^t \, dt = 365 \cdot ADTT \cdot \frac{(1+\alpha)^t - 1}{\ln(1+\alpha)}$$ (2-17)

where t is the number of years, subscript tr means trucks only, and α is the traffic increase rate per year.

29

The RRCs of the two lanes corresponding to the two traffic increase rates 3% and 5% are also shown in Fig. 2-4. Since there are only minor changes compared with the case without traffic increase, the progressive deterioration functions were assumed the same as the case with 0% traffic increase as shown in Eqs. (2-16) and (2-17).

2.5 Fatigue Reliability Assessment

2.5.1. Equivalent Stress Range

When vehicles travel along bridges, vehicle induced vibrations may generate stress ranges and fatigue damages may accumulate at bridge components, which could lead to bridge failures. Since only one truck is assumed to travel on the bridge at one time in the present study, the total stress history of the bridge can be simplified as a combination of the stress histories induced by vehicles with random speed. At each simulation, the road profile is generated randomly considering the progressive deterioration of the road surface condition.

Stress ranges are calculated case by case for different vehicle speeds and different road roughness conditions. Cycle-counting method, such as rainflow cycle-counting method, is used to process irregular stress histories and store the data. The total number of cycles, n_{tc}, is obtained after a stress range cut-off level is defined. Applicable cut-off levels for stress ranges are suggested in a range from 3.45 MPa (0.5ksi) to 33% the constant amplitude fatigue limit (CAFL) (Kwon and Frangopol 2010). The equivalent stress ranges are defined as the constant-amplitude stress range that can yield the same fatigue life as the variable-amplitude stress range for a structural detail. According to Miner's rule, the accumulated damage is

$$D(t) = \sum_i \frac{n_i}{N_i} = \frac{n_{tc}}{N} \tag{2-18}$$

where n_i is number of observations in the predefined stress-range bin S_{ri}, N_i is the number of cycles to failure corresponding to the predefined stress-range bin; n_{tc} is the total number of stress cycles and N is the number of cycles to failure under an equivalent constant amplitude loading (Kwon and Frangopol 2010):

$$N = A \cdot S_{re}^{-m} \tag{2-19}$$

where S_{re} is the equivalent stress range and A is the detail constant taken from Table 6.6.1.2.5-1 in AASHTO LRFD bridge design specifications (AASHTO, 2007). Either using the Miner's rule or Linear Elastic Fracture Mechanics (LEFM) approach, the equivalent stress range for the whole design life is obtained through the following equation (Chung 2006):

$$S_{re} = \left(\sum_{i=1}^{n} \alpha_i \cdot S_{ri}^m \right)^{1/m} \tag{2-20}$$

where α_i is the occurrence frequency of the stress-range bin, n is the total numbers of the stress-range bin and m is the material constant that could be assumed as 3.0 for all fatigue categories (Keating and Fisher 1986).

Since each truck passage might induce multiple stress cycles, two correlated parameters were essential to calculate the fatigue damages done by each truck passage, i.e. the equivalent stress range and numbers of cycle per truck passage. In the present study, a new single parameter, S_w, is introduced for simplifications to coalesce the two parameters on a basis of equivalent fatigue damage; namely, the fatigue damage of multiple stress cycles is the same as that of a single stress cycle of S_w. For truck passage j, the revised equivalent stress range is defined and derived as:

$$S_w^j = \left(N_c^j\right)^{1/m} \cdot S_{re}^j \tag{2-21}$$

where N_c^j is the number of stress cycles due to the jth truck passage, and S_{re}^j is the equivalent stress range of the stress cycles by the jth truck.

2.5.2. Limit State Function

When $D(t)$ is 1, the structure approaches to fatigue failure based on the Miner's rule. Correspondingly, the limit state function (LSF) is defined as (Nyman and Moses 1985):

$$g(X) = D_f - D(t) \tag{2-22}$$

where D_f is the damage to cause failure and is treated as a random variable with a mean value of 1; $D(t)$ is the accumulated damage at time t; and g is a failure function such that $g<0$ implies a fatigue failure.

In the preceding parts, the typical representative vehicle speed ranges (i.e. vehicle speed from 10m/s (22.4 mph) to 60m/s (134.4mph)), lane numbers (i.e. the fast lane or the slow lane) and road roughness conditions (i.e. from very good to very poor) have been defined. The overall fatigue damages are a summation of damages done by the trucks under all vehicle speed ranges, lane numbers and road roughness conditions. The accumulated damage $D(t)$ is:

$$D(t) = \sum_j \frac{n_{truck}^j \cdot N_c}{A \cdot \left(S_{re}^j\right)^{-m}} = \sum_j n_{truck}^j \cdot \left(S_w^j\right)^m A^{-1} = n_{tr} \cdot A^{-1} \cdot \sum_j \left(p_j\right) \cdot \left(S_w^j\right)^m \tag{2-23}$$

where, p_j means the probability of case j, and here case j is defined as a combination of vehicle speed, road roughness condition and lane numbers. Accordingly, a combination of the six vehicle speed ranges, five road roughness conditions and two lane numbers leads to 60 cases.

The overall equivalent stress range at time t can be obtained from Eqs. (2-19), (2-20) and (2-24) as:

$$S_{re} = \left(\frac{D(t) \cdot A}{n_{tr}}\right)^{1/m} = \left(\frac{\sum_j n_{truck}^j \cdot \left(S_w^j\right)^m}{\sum_j n_{truck}^j}\right)^{1/m} \tag{2-24}$$

The accumulated number of truck passage, n_{tr}, and the accumulated number of stress cycles, n_{tc} are obtained by the following equations, respectively:

$$n_{tr} = \sum_j n_{truck}^j \tag{2-25}$$

$$n_{tc} = \sum_j \left(n_{truck}^j \cdot N_c^j\right) \tag{2-26}$$

where, n_{truck}^j is the number of trucks corresponding to S_{re}^j.

2.5.3. Parameter Database

All the random variables for predicting fatigue reliabilities are listed in Table 2-4 including their distribution types, mean values, coefficient of variations (COVs) and descriptions.

31

Table 2-4 Summary of LSF parameters

Par.	Mean	COV	Distribution	Description
D_f	1.0	0.15	Lognormal	Damage to cause failure
ADTT	2000		Deterministic	ADTT in fatigue life
N_c	Calculated			Number of cycles per truck passage
t	75		Deterministic	Total fatigue life in years
A	7.83×10^{10}	0.34	Lognormal	Detail constant
m	3.0		Deterministic	Slope constant
S_w	Calculated	Calculated	Lognormal	Revised equivalent stress range
V	42.9mph (19.1m/s)	0.4	normal	Vehicle speed

D_f, the accumulated damage at failure, is considered as a random variable. Its mean and COV value is assumed as 1.0 and 0.15, respectively. The COV value are chosen to ensure that 95% of variable amplitude loading tests have a life within 70-130% (± 2 sigma) of the Miner's rule prediction (Nyman and Moses 1985).

The ADTT for trucks often varies greatly for bridges at different sites and causes the variations of the estimated fatigue life and fatigue reliability. The ADTT can be calculated and predicted from filed monitoring data. As discussed earlier, the ADTT for the trucks HS20-44 is assumed as a deterministic parameter that equals to 2000 in the first year. The ADTT number might remain the same or increase if a traffic increase is considered (i.e. $\alpha = 0\%$, 3% or 5%).

The present study is concerned with the fatigue cracks that may develop at the longitudinal welds located at the conjunction of the web and the bottom flange at the mid-span. In section 6.6 of AASHTO LRFD bridge design specifications, this type of fatigue-prone detail falls into Category B (AASHTO 2007) and the fatigue detail coefficient A can be obtained directly from the table in the specifications. When A is assumed to follow lognormal distribution, the mean and COV value can be calculated based on the test results of welded bridge details. Based on the tests performed by Keating and Fisher (1986), the mean and COV is calculated as 7.83×10^{10} and 0.34.

S_w, the revised equivalent stress range, is calculated for given combinations of vehicle speed and road roughness condition. Since the road profile is randomly generated for a given road roughness coefficient, the revised equivalent stress range might be different for each randomly generated road profile and might follow a certain distribution such as normal or lognormal distribution. In the present study, the chi-square goodness-of-fit test is used to check the distribution type of the parameter S_w for each combination of vehicle speed and road roughness condition. Therefore, when evaluating LSF, the revised equivalent stress range can be calculated based on the randomly generated vehicle speeds and road profiles.

Based on the assumption that all the variables (i.e. A, D_f and S_{wj}) follow a certain distribution, the fatigue reliability index is obtained using the method in the literature (Estes and Frangopol 1998). Based on their method, an arbitral initial design point can be chosen and the solving process for the complex equation of $g()=0$ can be avoided. After several iterations, convergence can be achieved without forcing every design points to fall on the original failure surface.

2.6 Results and Discussions

2.6.1. Cycles per Truck Passage

In section 6.6 of AASHTO LRFD bridge design specifications (AASHTO 2007), the number of cycles per truck passage is offered directly in Table 6.6.1.2.5-2. Regardless vehicle speed and the road surface condition, the number of cycles per truck passage for the bridge prototype is two, namely only the two stress cycles are assumed to cause fatigue damages. However, Albrecht and Friedland (1979) and Fisher et al (1983) indicated that fatigue cracks developed even though the equivalent stress range was well below the constant amplitude fatigue limit. In the real scenarios for predicting fatigue damages, a stress threshold needs to be defined initially to include the fatigue damages from the stress range cycles that are below the constant amplitude fatigue limit. Therefore, three values of threshold, i.e. 3.45Mpa (0.5ksi), 13.8Mpa (2ksi) and 34.5Mpa (5ksi), were chosen in the present study. These values fall into the range from 3.45Mpa (0.5ksi) to the 33% CAFL value as suggested by Kwon and Frangopol (2010). Since the road profile is randomly generated based on road condition indicator (i.e. from very good to very poor), the stress ranges and number of cycles are different. For each group of cases with the same vehicle speed and road roughness condition, more than twenty numerical simulations are carried out to obtain the mean and standard deviation of the number of cycles per truck passage and revised equivalent stress range. The mean values of the number of cycles per truck passage for the three thresholds were shown in Fig. 2-5 from (a) to (c), respectively. It is noteworthy that 3.45Mpa (0.5ksi) is a typically used threshold value for data analysis on the stress range obtained from field monitoring and is acceptable compared with the steel yield strength of 50ksi. However, the stress range cut-off would influence the numbers of cycles per truck passage but would not influence the distribution of equivalent stress range and the fatigue reliability and fatigue life since the stress ranges under 3.45Mpa (0.5ksi) only contribute a neglectable magnitude on the revised equivalent stress range.

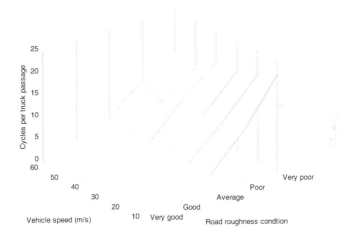

(a) Threshold = 3.45Mpa (0.5ksi)

Fig. 2-5.   Cycles per truck passage

(Fig. 2-5 continued)

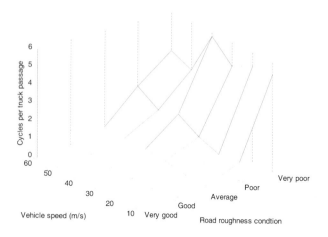

(b) Threshold = 13.8Mpa (2ksi)

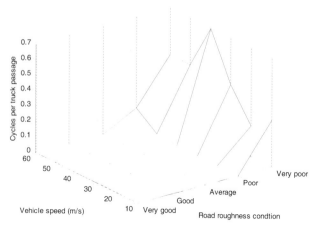

(c) Threshold = 34.5Mpa (5ksi)

As shown in Fig. 2-5(a), when the threshold value is 3.45Mpa (0.5ksi), the number of cycles per truck passage is between one and two when the road roughness condition is very good or good regardless the vehicle speed. However, the cycle numbers increase greatly as the road roughness condition deteriorates from average to very poor. At certain combinations of vehicle speeds and road roughness conditions, the numbers of cycles increase up to twenties. Generally, a more deteriorated road roughness condition leads to a larger number of cycles.

When the threshold value increases to 13.8Mpa (2ksi), as shown in Fig. 2-5(b), a great number of stress ranges is filtered out, which results a great decrease of stress cycles per truck passage.

When the road surface condition is from very good to poor, the number of cycles ranges from zero to two in most cases with only one exception case when the vehicle speed is 60m/s (134.4mph) under poor road roughness condition. When the road roughness condition is very good or good, most of the stress ranges are below the threshold that leads to a zero cycle count at several vehicle speeds. When the threshold increases to 34.5Mpa (5ksi), the number of cycles is larger than zero but less than 0.5 only at the vehicle speeds under very poor road roughness condition and two vehicle speeds under poor road roughness condition as shown in Fig. 2-5(c). Most cases do not have stress ranges larger than the threshold.

Based on the results of the present study, the cycles per truck passage cannot be treated as a constant value since they range from zero to twenties under different stress threshold values, different vehicle speeds and different road roughness conditions. In section 6.6 of AASHTO LRFD bridge design specifications, the cycle per truck passage is a constant value of one or two for most bridge components, which might greatly underestimate fatigue damages when existing bridges have deteriorated road roughness conditions.

2.6.2. Distributions of S_w

Same road roughness coefficient corresponds to numerous randomly generated road profiles and might lead to different stress magnitudes and numbers of stress cycles. In order to include the random effects of the road profile, it is necessary to check the distribution type of the revised equivalent stress range. The calculated stress ranges are used as samples for a fit-of-goodness test, such as Chi-Square test. In the Chi-Square test used in the present study, fifty road profile samples are generated for a given combination of vehicle speed and road roughness condition. Based on Sturges' rule, the number of intervals is seven for a 50-data bin and the degree of freedom is four. If a 5% significance level is chosen, the test limit for the Chi-Square test is calculated as $C_{1-\alpha,f} = C_{0.95,4} = 9.488$. In the present study, only six out of the total thirty groups of cases are employed as demonstrations to verify the distribution type of the revised equivalent stress range S_w in order to save calculation cost. The six groups of cases are the combinations of the road roughness condition from poor to very poor and vehicle speed of 40m/s (89.6 mph), 50m/s (112.0 mph) and 60m/s (134.4mph). Therefore, six groups of cases (altogether 300 cases) are checked for the distribution type.

When the threshold is 3.45Mpa (0.5ksi), the Chi-Square tests of revised equivalent stress range S_w for normal and lognormal distributions are listed in Table 2-5. It suggests that both normal and lognormal are acceptable distribution types for the revised equivalent stress range. When the threshold increases to 13.8Mpa (2ksi), a zero revised equivalent stress range was found in three cases out of the 300 cases, which greatly affects the distribution especially for the lognormal distribution. A star subscript is attached to the revised equivalent stress range when zero revised equivalent stress ranges are found as shown in Table 2-6. For other groups with smaller vehicle speeds and better road conditions, more zero revised equivalent stress range are found. When the threshold increases to 34.5Mpa (5 ksi), zero revised equivalent stress range is found almost in all groups of cases. Accordingly, the distribution type for the revised equivalent stress range cannot be assumed as normal or lognormal when the thresholds are 13.8MPa (2ksi) or 34.5Mpa (5ksi). It is noteworthy that no zero revised equivalent stress ranges are found when the threshold is 3.45Mpa (0.5ksi). In the present study, the threshold is chosen as 3.45Mpa (0.5ksi) and the revised equivalent stress range is assumed to follow a lognormal distribution in each combination of road roughness condition and vehicle speed. The mean and COV values of revised equivalent stress ranges are listed in Table 2-7.

Table 2-5 Chi-Square test for the revised equivalent stress range S_w
threshold =3.45Mpa (0.5 ksi)

U_{ve} / Roughness	poor		very poor	
	Normal	Logn	Normal	Logn
40m/s (89.6mph)	7.5	4.8	2.7	5.0
50m/s (112mph)	3.5	6.0	2.9	2.0
60m/s (134.4 mph)	6.5	5.5	3.5	3.3

Table 2-6 Chi-Square test for the revised equivalent stress range S_w
threshold =13.8 Mpa (2.0 ksi)

U_{ve} / Roughness	poor		very poor	
	Normal	Logn	Normal	Logn
40m/s (89.6mph)	6.2	2.3	4.5	3.6
50m/s (112mph)	5.5*	9.8*	0.8*	23.1*
60m/s (134.4 mph)	7.1	8.8	0.3*	41.1*

Table 2-7 Mean and Coefficient of Variation of S_w (threshold =3.45Mpa(0.5 ksi))

U_{ve} / Roughness	very good		good		average		poor		very poor	
	Mean	Cov	Mean	Cov	Mean	Cov	Mean	Cov	Mean	Cov
10m/s	1.48	0.03	1.62	0.06	1.93	0.13	2.95	0.16	5.84	0.24
20m/s	1.50	0.03	1.62	0.04	1.97	0.10	2.89	0.17	5.57	0.23
30m/s	1.54	0.04	1.62	0.06	2.19	0.10	3.37	0.19	6.02	0.25
40m/s	1.59	0.04	1.72	0.07	2.17	0.21	3.57	0.28	5.47	0.21
50m/s	1.59	0.08	1.74	0.14	2.28	0.22	3.49	0.24	5.66	0.33
60m/s	1.66	0.08	1.80	0.11	2.46	0.25	4.38	0.34	5.71	0.33

If the maximum stress range exceeds the corresponding CAFL, the structural detail may experience finite fatigue life and the structural component might have fatigue failure. However, before determining whether the structure has a finite fatigue life or not, the probability of the number of cycles exceeding the CAFL has to be defined first. In a conservative manner, the damage-causing frequency limit was set as 0.01% (Kwon and Frangopol 2010). Based on the assumption of the distribution of S_w and the mean value of the numbers of per truck passage, the frequency of number of cycles exceeding CAFL was found below the limit of 0.01% for all cases and all the cases were theoretically expected an infinite fatigue life. However, based on the assumed S-N curve in Eq. (2-20), the fatigue life and reliability index associated with the number of cycles corresponding to the equivalent stress range still can be estimated.

2.6.3. Fatigue Reliability Assessment and Prediction

In order to investigate the effects of the vehicle speed and road roughness condition on the fatigue reliability, the fatigue reliability are calculated for all the 30 combinations of 6 vehicle speeds from 10m/s (22.4 mph) to 60m/s (134.4mph) and 5 road roughness conditions from very good to very poor when the threshold is 3.45Mpa (0.5ksi). The LSF for a given combination of vehicle speed and road roughness condition is simplified as:

$$g(X) = D_f - \frac{n_{tr} \cdot S_w^m}{A} \tag{2-27}$$

Based on the LSF, the fatigue reliability index, β, can be derived, assuming that all random variables (i.e. A, D_f and S_w) are lognormal, as follows:

$$\beta = \frac{\lambda_{D_f} + \lambda_A - \left(m \cdot \lambda_{S_w} + \ln n_{tr} \right)}{\sqrt{\zeta_{D_f}^2 + \zeta_A^2 + (m \cdot \zeta_{S_w})^2}} \tag{2-28}$$

where $\lambda_{D_f}, \lambda_A, \lambda_{S_w}$ and $\zeta_{D_f}, \zeta_A, \zeta_{S_w}$ denote the mean value and standard deviation of $\ln(A)$, $\ln(D_f)$ and $\ln(S_w)$, respectively.

Fatigue reliability indices for the three vehicle increase rates (i.e. α=0%, 3% and 5%) are listed in Fig. 2-6 (a), (b) and (c). Generally, the fatigue reliability indices are found to decrease with the increase of vehicle speed and road roughness coefficient. If a 5% failure probability, i.e., a 95% survival probability is assumed, the corresponding reliability index is 1.65 (Kwon and Frangopol 2010). When the vehicle increase rate is 0%, the reliability index in all the thirty cases is larger than the target index of 1.65. Accordingly, the survival probability of all the thirty cases is larger than 95%. When the vehicle increase rate is 3% or 5%, seven or eight cases are found with a fatigue reliability index less than 1.65, i.e., the probability of fatigue failure for the bridge is larger than 5% during its 75 years life. In addition, when the vehicle increase rate is 5% and the road roughness condition is very poor, the reliability index is negative, which suggests the bridge would be more likely to suffer fatigue failure than to survive in its 75 years life. Based on the assumed target reliability index 1.65, the predicted fatigue life is shown in Fig. 2-7, which clearly indicates the effects of the vehicle speed and the road surface condition on the fatigue lives. In general, the higher vehicle speed, the smaller reliability index and the higher probability of failure the structure will have in most cases. The road condition makes great changes to the reliability index and results in a change from zero to more than 10. The change in the reliability index due to the vehicle speed was found to be less but still cannot be neglected. For example, at average road condition and 3% vehicle increase rate, the fatigue reliability varies from 4 to 8.2 (See Fig. 2-6(b)). The deteriorated road surface seems to accelerate fatigue damages due to the dynamic effects from vehicles, which implies the importance of road surface maintenances for existing bridges.

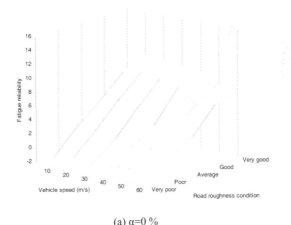

(a) α=0 %

Fig. 2-6.   Fatigue reliability index for given vehicle speed and road roughness condition

37

(Fig. 2-6 continued)

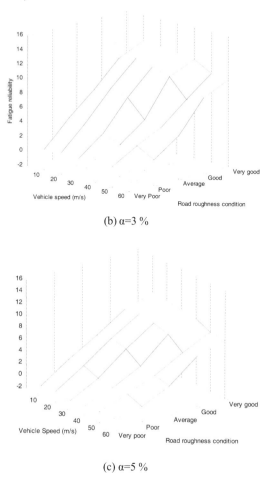

(b) α=3 %

(c) α=5 %

Table 2-8 Fatigue reliability index and Fatigue life (threshold =3.45Mpa(0.5 ksi))

α	β (for Fatigue life of 75 years)	Fatigue life for target β=1.65 (unit: years)
0 %	6.5	1906
3 %	6.0	136
5 %	4.4	93

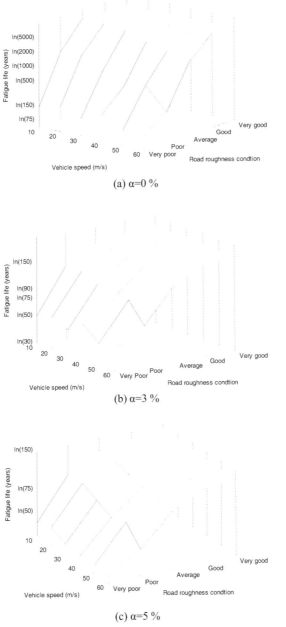

(a) α=0 %

(b) α=3 %

(c) α=5 %

Fig. 2-7.　Fatigue life for given vehicle speed and road roughness condition

In the real circumstances, vehicle speeds vary for different trucks and road surface conditions deteriorated with time. By treating the vehicle speed and road roughness as random variables as discussed earlier, the fatigue reliability index is calculated and listed in Table 2-8 for different traffic increase rate, i.e., $\alpha = 0\%$, 3% and 5%. The fatigue life corresponding to a target fatigue reliability of 1.65 is also presented in the table. As expected, the predicted fatigue life is longer than that if we assume the road roughness is poor or very poor and shorter than that if we assume the road roughness is good or very good (see Figs. 2-5 and 2-6). For the current modeling of the vehicle speed and road surface deteriorations, the fatigue life of the bridge components is comparable with the case with a 60m/s (134.4mph) vehicle speed and an average road-roughness condition.

2.7 Conclusions

This paper presents an approach for fatigue reliability assessment of existing bridges considering the random effects of vehicle speeds and deteriorating road roughness conditions of bridge decks. In the present study, fatigue reliability assessment of a short span slab-on-girder bridge under three-axle trucks are carried out based on the stress history obtained from the 3D bridge-vehicle interaction simulations in the time domain.

After setting up the limit state function with several random variables (including fatigue damages to cause failure, vehicle speeds, road roughness conditions, the revised equivalent stress ranges and the constant amplitude fatigue thresholds), fatigue reliability of the structural details is attained. Chi-square test is used in the present study and the revised equivalent stress range is found to follow a lognormal distribution when the threshold of the stress ranges is 3.45Mpa (0.5ksi) or below. In addition, lognormal distribution is unacceptable for the stress ranges when the threshold increases to 13.8Mpa (2ksi) or 34.5Mpa (5ksi). Future fatigue life can also be calculated when the target reliability index is defined, such as 1.65. From the present study, the following conclusions are drawn:

1. The vehicle speed affects the fatigue reliability and fatigue life of the bridge components. In most cases, a higher vehicle speed induces a larger stress range and a larger number of cycles per truck passage. Accordingly, the fatigue reliability decreases with the increase of vehicle speed.

2. The road roughness condition influences the fatigue reliability of the bridge components. Generally, the more deteriorated road condition induces larger stress ranges and larger numbers of stress cycles for each truck passage, which leads to a smaller fatigue reliability index.

3. The cut-off threshold of stress ranges has a significant effect on stress contribution and further study is needed to decide a rational value. In the present study, the revised equivalent stress range follows lognormal distribution when the stress range threshold is 3.45Mpa (0.5ksi) and below. It does not follow a lognormal distribution when the threshold increases to 13.8Mpa (2ksi) or 34.5Mpa (5ksi).

4. With the increase of traffic increase rates, the fatigue reliability drops and the fatigue life reduces significantly.

In the present study, only the effect of the design truck, i.e. a three-axle truck, is deliberated. Based on the WIM data, a small percentage of trucks are heavier than the design vehicle. Based on the present study, the heavier trucks most likely do more damages to the road surface and might introduce larger stress ranges and more stress range cycles. As a result, the small percentage of heavy trucks might induce a large drop of the fatigue reliability index and fatigue life, accordingly. Since the real trucks vary in axle numbers, distances and weights, it is necessary to propose a more general and convenient approach to obtain the stress ranges considering the interactions between vehicles and bridges. A suggested table similar to the Table 6.6.1.2.5-2 in AASHTO LRFD bridge design specifications is also needed to simplify the calculation approaches in the present study for practical applications.

2.8 References

American Association of State Highway and Transportation Officials (AASHTO). (2007). LRFD bridge design specifications, Washington, DC.

Albrecht, P., and Friedland, I. M. (1979). "Fatigue-Limit Effect on Variable-Amplitude Fatigue of Stiffeners." *Journal of the Structural Division*, 105(12), 2657-2675.

Blejwas, T. E., Feng, C. C., and Ayre, R. S. (1979). "Dynamic interaction of moving vehicles and structures." *Journal of Sound and Vibration*, 67, 513-521.

Cai, C. S., and Chen, S. R. (2004). "Framework of vehicle-bridge-wind dynamic analysis." *Journal of Wind Engineering and Industrial Aerodynamics*, 92(7-8), 579-607.

Cai, C. S., Shi, X. M., Araujo, M., and Chen, S. R. (2007). "Effect of approach span condition on vehicle-induced dynamic response of slab-on-girder road bridges." *Engineering Structures*, 29(12), 3210-3226.

Cheng, Y. B., Feng, M. Q., and Tan, C.-A. (2006). "Modeling of traffic excitation for system identification of bridge structures." *Computer Aided Civil & Infrastructure Engineering*, 21(1), 57.

Chung, H.-Y., Manuel, L., and Frank, K. H. (2006). "Optimal Inspection Scheduling of Steel Bridges Using Nondestructive Testing Techniques." Journal of Bridge Engineering, 11(3), 305-319.

Deng, L., and Cai, C. S. (2010). "Development of dynamic impact factor for performance evaluation of existing multi-girder concrete bridges." *Engineering Structures*, 32(1), 21-31.

Ditlevsen, O., and Madsen, H. O. (1994). "Stochastic Vehicle-Queue-Load Model for Large Bridges." *Journal of Engineering Mechanics*, 120(9), 1829-1847.

Dodds, C. J., and Robson, J. D. (1973). "The Description of Road Surface Roughness." *Journal of Sound and Vibration*, 31(2), 175-183.

Donnell, E. T., Hines, S. C., Mahoney, K. M., Porter, R. J., and McGee, H. (2009). "Speed Concepts: Informational Guide." U.S. Department of Transportation and Federal Highway Administration. Publication No. FHWA-SA-10-001.

Estes, A. C., and Frangopol, D. M. (1998). "RELSYS: A computer program for structural system reliability." *Structural Engineering and Mechanics*, 6(8), 901.

Fisher, J. W., Mertz, D. R., and Zhong, A. (1983). "Steel bridge members under variable amplitude long life fatigue loading." NCHRP Report 267, Transportation Research Board, Washington, D.C..

Guo, W. H., and Xu, Y. L. (2001). "Fully computerized approach to study cable-stayed bridge-vehicle interaction." *Journal of Sound and Vibration*, 248(4), 745-761.

Honda, H., Kajikawa, Y., and Kobori, T. (1982). "SPECTRA OF ROAD SURFACE ROUGHNESS ON BRIDGES." *Journal of the Structural Division*, 108(ST-9), 1956-1966.

International Standard Organization. (1995). "Mechanical vibration - Road surface profiles - Reporting of measured data." Geneva.

Keating, P. B., and Fisher, J. W. (1986). "Evaluation of Fatigue Tests and Design Criteria on Welded Details." NCHRP Report 286, Transportation Research Board, Washington, D.C.

Kwon, K., and Frangopol, D. M. (2010). "Bridge fatigue reliability assessment using probability density functions of equivalent stress range based on field monitoring data." *International Journal of Fatigue*, 32(8), 1221-1232.

Laman, J. A., and Nowak, A. S. (1996). "Fatigue-Load Models for Girder Bridges." *Journal of Structural Engineering*, 122(7), 726-733.

Moses, F. (2001). "Calibration of load factors for LRFR Bridge." NCHRP Report 454, Transportation Research Board, Washington, D.C.

Nyman, W. E., and Moses, F. (1985). "Calibration of Bridge Fatigue Design Model." *Journal of Structural Engineering*, 111(6), 1251-1266.

O'Connor, A., and O'Brien, E. J. (2005). "Traffic load modeling and factors influencing the accuracy of predicted extremes." *Canadian Journal of Civil Engineering*, 32(1), 270-278.

Paterson, W. D. O. (1986). "International Roughness Index: Relationship to Other Measures of Roughness and Riding Quality." Transportation Research Record 1084, Washington, D.C.

Salane, H.J. and Baldwin, J.W. (1990) "Changes in modal parameters of a bridge during fatigue testing." *Experimental Mechanics*, 30(2), 109-133

Sayers, M. W., and Karamihas, S. M. (1998). The Little Book of Profiling - Basic Information about Measuring and Interpreting Road Profiles.

Shi, X., Cai, C. S., and Chen, S. (2008). "Vehicle Induced Dynamic Behavior of Short-Span Slab Bridges Considering Effect of Approach Slab Condition." *Journal of Bridge Engineering*, 13(1), 83-92.

Shiyab, A. M. S. H. (2007). "Optimum Use of the Flexible Pavement Condition Indicators in Pavement Management System." Ph.D Dissertation, Curtin University of Technology.

Timoshenko, S., Young, D. H., and Weaver, W. (1974). *Vibration problems in engineering*, Wiley, New York.

TxDOT. (2006). "Procedures for Establishing Speed Zones." Texas Department of Transportation.

Wang, T.-L., and Huang, D. (1992). "Computer modeling analysis in bridge evaluation." Florida Department of Transportation, Tallahassee, FL.

CHAPTER 3   RELIABILITY BASED DYNAMIC AMPLIFICATION FACTOR ON STRESS RANGES FOR FATIGUE DESIGN OF EXISTING BRIDGES

3.1 Introduction

Fatigue is one of the main forms of structural damage and failure modes caused by repeated dynamic load effects. A great deal of attentions has been paid to the deterioration of the civil infrastructure, which might induce the collapse and fracture of structures. Procedures for applying fatigue reliability analysis of structures to reassess the fatigue life of existing structures were summarized by Byers et al. (1997a, 1997b). Once the current condition of a structure has been assessed, the remaining service life can be estimated based on the variable stress range histories using either fatigue life methods or a fracture mechanics approach (Cheung and Li 2003, Chung et al 2006, Pipinato et al 2011, Zhang and Cai 2011). Such variable stress ranges are induced by the moving vehicles for highway bridges. In 1982, the ASCE Committee on Fatigue and Fracture Reliability discussed possible use of probabilistic distributions for fatigue analysis (ASCE 1982). Hereafter, several probability density functions (PDFs), for instance, Weibull, Beta, and Lognormal distributions, were used to estimate equivalent stress range (Chung 2004, Pourzeynali and Datta 2005, Kwon and Frangopol 2010). Certain actions can be taken based on the results from the fatigue reliability analysis, for instance, repairing the structure, replacing the structure or changing the operation of the structure (Byers et al. 1997a).

Moving vehicles on a bridge usually generates greater deflections and stresses in the structure than those caused by the same vehicle loads applied statically. In many specifications including AASHTO LRFD (2010), the dynamic load allowance (IM) is defined as an increment to be applied to the static wheel load to account for dynamic impact from moving vehicles. Therefore, the maximum dynamic response of the moving vehicles can be obtained (Paultre et al. 1992) as:

$$R_{dyn} = R_{sta} \cdot \left(1 + IM/100\right) \tag{3-1}$$

where R_{sta} is the maximum static response, IM/100 is the dynamic amplifications (DA), and $\left(1 + IM/100\right)$ is the dynamic amplification factor (DAF) for the bridge. For example, a DAF value of 1.15 corresponds to a DA of 0.15 and an IM of 15%. The IM adopted for fatigue and fracture limit state and all the other limit states for bridge components except for the joints are 15% and 33% in AASHTO LRFD (2010). Billing (1984) presented the equations for computing dynamic amplifications (DA) when dealing with the bridge responses from vehicle passing:

$$
\begin{aligned}
DA &= (DPR - SPR)/SPX & &\textit{for the positive region} \\
DA &= Max(DNB - SNB, DNA - SNA)/SPX & &\textit{for the negative region} \\
DA &= RES/SPX & &\textit{for the residual region}
\end{aligned}
\tag{3-2}
$$

where SPX is the largest static response and all the other variables are defined in Fig. 3-1, DPR and SPR are the maximum dynamic positive response and maximum static positive response in the middle of the main span, DNA and SNA is the minimum dynamic negative response and minimum static negative response in the right side span of "negative after", DNB and SNB is the minimum dynamic negative response and minimum static negative response in the left side span of "negative before". The DA for the bridge is the largest DA, which is used for the calculations of deflections, moments, shears and stresses to account for the dynamic load effects.

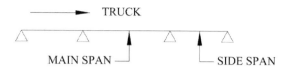

TRUCK

MAIN SPAN —┘   └— SIDE SPAN

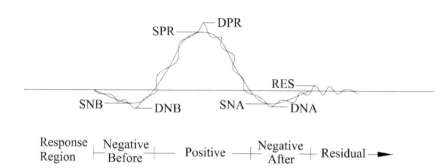

DPR
SPR

RES

SNB   DNB   SNA   DNA

Response | Negative | Positive | Negative | Residual ➤
Region | Before | | After |

Fig. 3-1 Schematic of bridge responses (adapted from Billing, 1984)

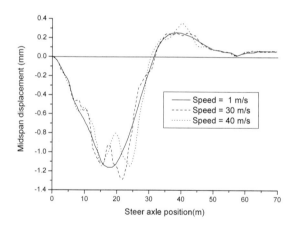

Fig. 3-2 Calculated dynamic responses of a three-span prestressed concrete bridge to a four-axle vehicle (adapted from Green, 1993)

Due to varied dynamic amplification effects in different regions, the calculated live load stress ranges might not be correct if the DA is used for the fatigue design. Billing (1984) presented an example to illustrate the possible effects of neglecting the differences of the dynamic amplification on negative and positive regions. For example, if the responses shown in Fig. 3-1 are stresses, and SPX = 1.0, DPR=1.2, SPR=1.0, DNB=-0.5, SNB=-0.3, DA can be obtained as 0.2 from Eq. (3-1). The dynamic stress range obtained using DA is $1.3 \times (1+0.2) =1.56$, while the actual stress range is 1.7. Therefore, the underestimated negative region dynamic response could lead to a potential

44

overestimate in fatigue life of 29%, namely $(1.7/1.56)^3=1.29$. Green (1993) also presented a case as shown in Fig. 3-2 in which large dynamic responses can be observed while DA is approximately zero. In such cases, DA is not an effective measure of dynamic stress cycles that are important for the fatigue design.

While many parameters such as the first natural frequency of the bridge, vehicle speed, suspension systems of vehicles, and initial vehicle vibrations have effects on the DAF, the road surface profiles have a tremendous effect (Paultre et al. 1992, Park et al. 2005, Ashebo et al. 2007, Ding et al. 2009). The AASHTO C4.7.2.1 also indicates that the deck surface roughness is a major factor in vehicle/bridge interaction and it is difficult to estimate the long-term deck deterioration effects thereof at the bridge design stage. However, the IM value prescribed by the LRFD code is based on the numerical simulations that consider only an average road surface condition (Hwang and Nowak 1991). Numerical simulations and field testing results have shown that the IM values are underestimated especially on a poor road surface condition, a high vehicle speed or the combined conditions (Billing 1984; O'Connor and Pritchard 1985; Shi et al. 2008; Deng et al. 2011). Nevertheless, when one truck travels along the bridge, only one or two stress cycles are considered in the AASHTO LRFD code for most small and medium span bridges. In the present study, the term of "each truck passage" is used to illustrate the dynamic interactions of the bridge and one moving vehicle along the bridge. The analytical and experimental results on several bridges indicated clearly that more than one or two stress ranges could be induced by each truck passage (Agarwal and Billing 1990, Nassif et al. 2003, Zhang and Cai 2011). Since the DA only reflects the largest stress amplitude during one vehicle passing on the bridge, fatigue damages from the other stress cycles with varied stress ranges might be underestimated. Therefore, the fatigue damages from dynamic vehicle load might not be correct and it is necessary to propose an effective measure for the dynamic stress range cycles for the fatigue design.

In the present study, a reliability based dynamic amplification factor on stress ranges (DAFS) for fatigue design is proposed to include the fatigue damages from multiple stress range cycles due to each vehicle passage at varied vehicle speeds under various road conditions in the bridge's life cycle. The paper is organized as the following three main sections. In the first section, the process of stress range acquisition is detailed. After introducing the vehicle-bridge dynamic system, the principles for generating stochastic random road profiles and the parameters used for the vehicle-bridge dynamic system are introduced including the prototype of the vehicle and bridge, road conditions and vehicle speeds. In the second section, the dynamic amplification factor on stress ranges (DAFS) are defined and parametric study of the DAFS in life cycle is carried out. Based on the Miner's linear fatigue damage model, the fatigue damage accumulation can be achieved. On an equivalent fatigue damage basis, a revised equivalent stress range is defined to use one stress cycle to reflect the fatigue damages from multiple stress ranges with varied amplitudes. At a given target reliability index, a nominal live load stress range can be obtained for a given progressive road surface deterioration model and given distributions of the random parameters, including vehicle speed and type and annual traffic increase rate. DAFS is then defined as the ratio of the nominal live load stress range and the maximum static stress range. A parametric study on DAFS is carried out to analyze the effect from the progressive road surface deterioration model and distributions of the random parameters for a given bridge in its design life, including vehicle speed and type and annual traffic increase rate. In order to appreciate the difference of the proposed DAFS and traditional DAF, the results from six deterministic or probabilistic approaches related to the DAFS and DAF for fatigue life estimation are compared with each other in the third section.

3.2 Stress Range Acquisition

In this section, the principles of stress range acquisition are detailed. Based on the randomly generated road profiles and the parameters defined for the dynamic system, such as vehicle speed, the stress ranges with variable amplitude are obtained by solving the equations of motions for the vehicle-bridge dynamic system. The prototypes of the bridge and vehicle models used in the present study are introduced in this section, as well.

3.2.1. Vehicle-Bridge Dynamic System

In order to make an accurate estimation of fatigue life of existing bridges, it is necessary to predict a reasonable future stress range history due to various traffic loadings under various road surface conditions. Such data can be obtained either from on-site strain measurements or structural dynamic analysis of bridges. However, stress range spectra for bridges are strongly site-specific due to different vehicle types and speed distributions, road roughness conditions and bridge types (Laman and Nowak 1996). Therefore, it would be impossible to use the on-site measurement for every bridge or every concerned location of a bridge for a given traffic-loading pattern including the vehicle types, speeds and road surface conditions. Reasonable stress range data in various scenarios for bridge details can be provided by numerical simulations in a full vehicle-bridge coupled dynamic system (Guo and Xu 2001). The interactions between the bridge and vehicles are modeled as coupling forces between the tires and the road surface. The coupling forces were proven to be significantly affected by the vehicle speed and road roughness conditions and resulted in significant effects on the dynamic responses of short span bridges (Cai et al. 2011, Deng and Cai 2010; Shi et al. 2008). In order to include fatigue damages from the stress ranges with variable amplitudes associating with various vehicle speeds and progressively deteriorating road roughness conditions, a framework of fatigue reliability assessment for existing bridges was proposed by Zhang and Cai (2011).

In the present study, the vehicle is modeled as a combination of several rigid bodies connected by several axle mass blocks, springs, and damping devices (Cai and Chen 2004). The tires and suspension systems are idealized as linear elastic spring elements and dashpots. As a large number of degrees of freedom (DOF) are involved, the mode superposition technique is used to simplify the modeling procedure based on the obtained bridge mode shape and the corresponding natural circular frequencies.

The equation of motion for the vehicle and the bridge are listed in the following matrix form:

$$[M_v]\{\ddot{d}_v\}+[C_v]\{\dot{d}_v\}+[K_v]\{d_v\}=\{F_v^G\}+\{F_c\} \qquad (3\text{-}3)$$

$$[M_b]\{\ddot{d}_b\}+[C_b]\{\dot{d}_b\}+[K_b]\{d_b\}=\{F_b\} \qquad (3\text{-}4)$$

where $[M_v]$, the mass matrix, $[C_v]$, damping matrix and $[K_v]$, stiffness matrix are obtained by considering the equilibrium of the forces and moments of the system; $\{F_v^G\}$ is the self-weight of the vehicle; $\{F_c\}$ is the vector of wheel-road contact forces acting on the vehicle; $[M_b]$ is the mass matrix, $[C_b]$ is the damping matrix; $[K_b]$ is the stiffness matrix of the bridge; and $\{F_b\}$ is wheel-bridge contact forces on bridge and can be stated as a function of deformation of the vehicle's lower spring:

$$\{F_b\}=-\{F_c\}=[K_l]\{\Delta_l\}+[C_l]\{\dot{\Delta}_l\} \qquad (3\text{-}5)$$

where $[K_l]$ and $[C_l]$ are coefficients of vehicle lower spring and damper; and Δ_l is deformation of lower springs of vehicle. The relationship among vehicle-axle-suspension displacement Z_a, displacement of bridge at wheel-road contact points Z_b, deformation of lower springs of vehicle Δ_l, and road surface profile $r(x)$ is:

$$Z_a=Z_b+r(x)+\Delta_l \qquad (3\text{-}6)$$

$$\dot{Z}_a=\dot{Z}_b+\dot{r}(x)+\dot{\Delta}_l \qquad (3\text{-}7)$$

46

where $\dot{r}(x) = \left(dr(x) / dx \right) \cdot \left(dx / dt \right) = \left(dr(x) / dx \right) \cdot V(t)$ and $V(t)$ is the vehicle velocity.

Therefore, the contact force F_b and F_c between the vehicle and the bridge is:

$$\{F_b\} = -\{F_c\} = [K_l]\{Z_a - Z_b - r(x)\} + [C_l]\{\dot{Z}_a - \dot{Z}_b - \dot{r}(x)\} \tag{3-8}$$

After transforming the contact forces to equivalent nodal force and substituting them into Eqs. (3) and (4), the final equations of motion for the coupled system are as follows (Shi et al. 2008):

$$\begin{bmatrix} M_b & \\ & M_v \end{bmatrix} \begin{Bmatrix} \ddot{d}_b \\ \ddot{d}_v \end{Bmatrix} + \begin{bmatrix} C_b + C_{bb} & C_{bv} \\ C_{vb} & C_v \end{bmatrix} \begin{Bmatrix} \dot{d}_b \\ \dot{d}_v \end{Bmatrix} + \begin{bmatrix} K_b + K_{bb} & K_{bv} \\ K_{vb} & K_v \end{bmatrix} \begin{Bmatrix} d_b \\ d_v \end{Bmatrix} = \begin{Bmatrix} F_{br} \\ F_{vr} + F_v^G \end{Bmatrix} \tag{3-9}$$

The additional terms C_{bb}, C_{bv}, C_{vb}, K_{bb}, K_{bv}, K_{vb}, F_{br} and F_{vr} in Eq. (3-9) are due to the expansion of the contact force in comparison with Eqs. (3) and (4). When the vehicle is moving across the bridge, the bridge-vehicle contact points change with the vehicle position and the road roughness at the contact point. After obtaining the bridge dynamic response $\{d_b\}$, the stress vector can be obtained by:

$$[S] = [E][B]\{d_b\} \tag{3-10}$$

where $[E]$ is the stress-strain relationship matrix and is assumed to be constant over the element and $[B]$ is the strain-displacement relationship matrix assembled with x, y and z derivatives of the element shape functions.

3.2.2. Stochastic Random Road Profile

In the current AASHTO LRFD specifications (AASHTO LRFD 2010), the dynamic effects due to moving vehicles are attributed to two sources, namely, the hammering effect due to vehicle riding surface discontinuities, such as deck joints, cracks, potholes and delaminations, and dynamic response due to long undulations in the roadway pavement.

Based on the studies carried out by Dodds and Robson (1973) and Honda et al. (1982), the long undulations in the roadway pavement could be assumed as a zero-mean stationary Gaussian random process and it could be generated through an inverse Fourier transformation (Wang and Huang 1992):

$$r(x) = \sum_{k=1}^{N} \sqrt{2\phi(n_k)\Delta n} \cos(2\pi n_k x + \theta_k) \tag{3-11}$$

where θ_k is the random phase angle uniformly distributed from 0 to 2π; $\phi()$ is the power spectral density (PSD) function (m^3/cycle) for the road surface elevation; n_k is the wave number (cycle/m). The PSD functions for road surface roughness were developed by Dodds and Robson (1973), and three groups of road classes were defined with the values of roughness exponents ranging from 1.36 to 2.28 for motorways, principal roads, and minor roads. In order to simplify the description of road surface roughness, both of the two roughness exponents were assumed to have a value of two and the PSD function was simplified by Wang and Huang (1992) as:

$$\phi(n) = \phi(n_0)(\frac{n}{n_0})^{-2} \tag{3-12}$$

47

where $\phi(n)$ is the PSD function (m^3/cycle) for the road surface elevation; n is the spatial frequency (cycle/m); n_0 is the discontinuity frequency of $1/2\pi$ (cycle/m); and $\phi(n_0)$ is the road roughness coefficient (m^3/cycle) whose value is chosen depending on the road condition.

In order to include the progressive pavement damages due to traffic loads and environmental corrosions, a progressive road roughness deterioration model for the bridge deck surface is used (Zhang and Cai, 2011):

$$\phi_t(n_0) = 6.1972 \times 10^{-9} \times \exp\left\{\left[8.39 \times 10^{-6} \phi_0 e^{\eta t} + 263(1+SNC)^{-5}\left(CESAL\right)_t\right]/0.42808\right\} + 2 \times 10^{-6} \quad (3\text{-}13)$$

where ϕ_t is the road roughness coefficient at time t; ϕ_0 is the initial road roughness coefficient directly after completing the construction and before opening to traffic; t is the time in years; η is the environmental coefficient varying from 0.01 to 0.7 depending upon the dry or wet, freezing or non-freezing conditions; SNC is the structural number modified by sub grade strength and $(CESAL)_t$ is the estimated number of traffic in terms of AASHTO 18-kip (80kN) cumulative equivalent single axle load at time t in millions.

For the surface discontinuities that cause hammer effects, these irregularities, such as the uneven joints, the potholes and faulting (bumps), have a significant influence on bridge dynamic response and should be isolated and treated separately from such pseudo-random road surface profiles according to ISO (ISO 8606, 1995) and Cebon (1999). The local unevenness of expansion joints at the approach slab ends was found to increase the dynamic response of short span bridges. The discontinuities can be modeled with a step up or down for the faulting between approach slab and pavement and between bridge deck and approach slab (Green et al. 1997, Shi et al. 2008). According to US Federal Highway Administration (Miller and Bellinger, 2003), low, moderate and high severity potholes in the pavement are defined as 0.025m, 0.025-0.05m and more than 0.05m deep. In the present study, faulting of 0.038m is used to model the surface discontinuities. The discontinuities are assumed to be located at the entrance of the bridge. However, the vehicle is assumed traveling before entering the bridge. In the present study, the faulting values are assumed at both ends of the approach slab with the same values and the approach slab deflection and slope change in the approach slab was also used in the road profile to describe vehicle riding surface discontinuities (Shi et al. 2008, Cai et al. 2005). Therefore, a twofold road surface condition is used in the vehicle-bridge dynamic analysis to include the two sources for dynamic effects due to moving vehicles. Since most of the major road damages are expected to finish in one day, the default faulting day in each year is assumed as one.

3.2.3. Bridge and Vehicle Model

The short span bridges might be more vulnerable to suffer fatigue damages from variable dynamic stress ranges due to vehicle loads. To demonstrate the equivalent fatigue damage load methodology for bridges, a short span slab-on-girder bridge, a commonly used type of bridges in highways, is analyzed. The bridge is designed in accordance with AASHTO LRFD bridge design specifications (AASHTO LRFD 2010). The bridge has a span length of 12 m and a width of 13m, which accommodates two vehicle lanes traveling in the same direction. The concrete deck is 0.19m thick and the haunch is 40mm high. All of the six steel girders are W27×94 and have an even spacing of 2.3m as shown in Fig. 3-3. Two intermediate and two end cross-frames enable the girders to deflect more equally. In this bridge, a steel channel section, C15×33.9, is used as a cross-frame. The fundamental frequency of the bridge is 14.5 Hz. The damping ratio is assumed to be 0.02. As a demonstration, the present study focuses on the fatigue analysis at the longitudinal welds located at the conjunction of the web and the bottom flange at the mid-span as shown in Fig. 3-3.

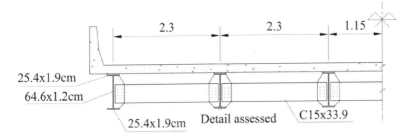

Fig. 3-3 Typical section of bridge (unit= meter)

In order to get the actual truckload spectra, weigh-in-motion (WIM) methodologies have been developed and are extensively used nationwide. Based on the data from WIM measurements, fifteen vehicle types are defined according to FHWA classification scheme "F". Types five, eight and nine, representing the typical trucks with axle numbers of two, three and five, are predominantly found according to traffic data in the WIM stations in Florida(Wang and Liu 2000). In the present study, their three-dimensional mathematic models are used and the average daily truck traffic for the truck with two, three and five axles is assumed to be 600, 400, and 1000. Due to the small length of the bridge, only one truck is assumed passing the whole bridge at one time. The distributions of the vehicle speed are assumed as the same for all the three types of vehicles.

The AASHTO H20-44, HS20-44 and 3S2 are used in the present study to represent the trucks with two, three and five axles as shown in Figs. 3-4 to 3-6, respectively. The geometry, mass distribution, damping, and stiffness of the tires and suspension systems of this truck are listed in Tables 3-1 to 3-3, respectively. It is noteworthy that the design live load for the prototype bridge is HS20-44 truck. The purpose of using the three types of trucks in the present study is to make a comparison and investigate their effects on DAFS. A 6 m long approach slab connecting the pavement and bridge deck is considered.

The dynamic displacement of bridges was found to be changing with the vehicle speed in the literature (Green 1990, Paultre et al. 1992, Cai and Chen 2004; Cai et al. 2007). Typically, the maximum speed limits posted in bridges or roads are based on the 85th percentile speed when adequate speed samples are available. The 85th percentile speed is a value that is used by many states and cities for establishing regulatory speed zones (Donnell et al. 2009; TxDOT 2006). Statistical techniques show that a normal distribution occurs when random samples of traffic are collected. This allows describing the vehicle speed conveniently with two characteristics, i.e. the mean and standard deviation. In the present study, the 85th percentile speed is approximated as the sum of the mean value and one standard deviation for simplification. In the normal design condition, the speed limit is assumed as 26.8m/s (60mph) and the coefficient of variance of vehicle speeds is assumed as 0.2.

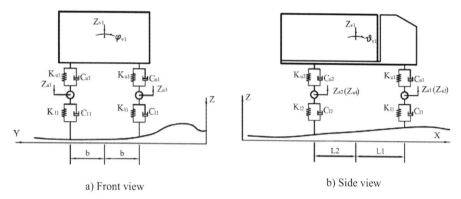

a) Front view b) Side view

Fig. 3-4 Vehicle model for two axles

Table 3-1 Major parameters of vehicle (2 axles)

Mass	truck body	15233 kg
	first axle suspension	725 kg
	second axle suspension	725 kg
Moment of inertia	Pitching, truck body1	19373 kg.m^2
	Rolling, truck body2	57690 kg.m^2
Spring Stiffness	Upper, 1st axle	242604 N/m
	Lower , 1st axle	875082 N/m
	Upper, 2nd axle	1903172 N/m
	Lower , 2nd axle	3503307 N/m
Damping coefficient	Upper, 1st axle	1314 N.s/m
	Lower , 1st axle	2000 N.s/m
	Upper, 2nd axle	7445 N.s/m
	Lower , 2nd axle	2000 N.s/m
Length	L1	3.41 m
	L2	0.85 m
	B	1.1 m

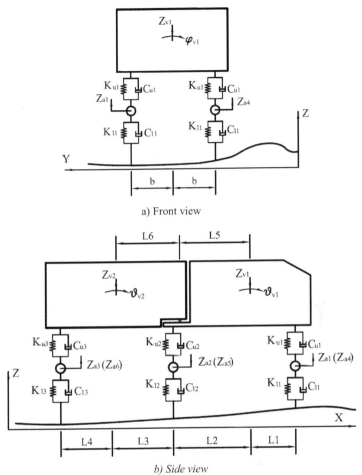

a) Front view

b) Side view

Fig. 3-5 Vehicle model for three axles

Table 3-2 Major parameters of vehicle (3 axles)

Mass	truck body 1	2612 kg
	truck body 2	26113 kg
	first axle suspension	490 kg
	second axle suspension	808 kg
	third axle suspension	653 kg
Moment of inertia	Pitching, truck body1	2022 kg.m2
	Pitching, truck body2	33153 kg.m2
	Rolling, truck body2	8544 kg.m2
	Rolling, truck body2	181216 kg.m2
Spring stiffness	Upper, 1st axle	242604 N/m
	Lower , 1st axle	875082 N/m
	Upper, 2nd axle	1903172 N/m
	Lower , 2nd axle	3503307 N/m
	Upper, 3rd axle	1969034 N/m
	Lower , 3rd axle	3507429 N/m
Damping coefficient	Upper, 1st axle	2190 N.s/m
	Lower , 1st axle	2000 N.s/m
	Upper, 2nd axle	7882 N.s/m
	Lower , 2nd axle	2000 N.s/m
	Upper, 3rd axle	7182 N.s/m
	Lower , 3rd axle	2000 N.s/m
Length	L1	1.698 m
	L2	2.569 m
	L3	1.984 m
	L4	2.283 m
	L5	2.215 m
	L6	2.338 m
	B	1.1 m

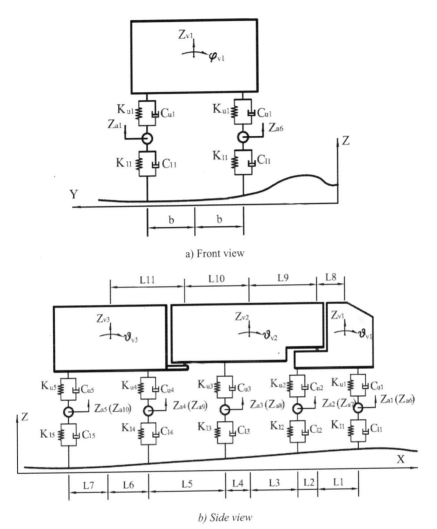

a) Front view

b) Side view

Fig. 3-6 Vehicle model for five axles

Table 3-3 Major parameters of vehicle (5 axles)

Mass	truck body 1	4956 kg
	truck body 2 & 3	20388 kg
	first axle suspension	297 kg
	2nd & 3rd axle suspension	892 kg
	4th & 5th axle suspension	1054 kg
Moment of inertia	Pitching, truck body1	3836 kg.m2
	Pitching, truck body2&3	20296 kg.m2
	Rolling, truck body1	12291 kg.m2
	Rolling, truck body2&3	333875 kg.m2
Spring stiffness	Upper, 1st axle	485208 N/m
	Lower, 1st axle	1402724 N/m
	Upper, 2nd & 3rd axle	1396068 N/m
	Lower, 2nd & 3rd axle	5610546 N/m
	Upper, 4th & 5th axle	1359634 N/m
	Lower, 4th & 5th axle	5610546 N/m
Damping coefficient	Upper, 1st axle	2400 N.s/m
	Lower, 1st axle	1600 N.s/m
	Upper, 2nd & 3rd axle	7214 N.s/m
	Lower, 2nd & 3rd axle	1600 N.s/m
	Upper, 4th & 5th axle	7574 N.s/m
	Lower, 4th & 5th axle	1600 N.s/m
Length	L1	3 m
	L2	5 m
	L3	1.64 m
	L4	3.36 m
	L5	2.0 m
	L6	3.055 m
	L7	1.945 m
	L8	2.4 m
	L9	1.64 m
	L10	3.36 m
	L11	5.05 m
	B	1.1 m

3.3 Dynamic Amplification Factor on Stress Ranges (DAFS)

In this section, the dynamic amplification factor on stress ranges (DAFS) is defined and a parametric study of the DAFS in life cycle is carried out. At first, the revised equivalent stress range is defined based on an equivalent fatigue damage basis. The acquired stress ranges from the last section can be used to calculate the revised equivalent stress range for given road surface condition and vehicle speed. After obtaining the nominal live load stress range, the DAFS can be obtained for a given road roughness condition, vehicle speed or for bridge's life cycle. At the end of this section, a parametric study is carried out to analyze the effect on DAFS from the progressive road surface deterioration model and distributions of the random parameters for a given bridge in its design life, including vehicle speed and type and annual traffic increase rate.

3.3.1. Revised Equivalent Stress Range

Since each truck passage might induce multiple stress cycles, two correlated parameters are essential to calculate the fatigue damages done by each truck passage, i.e. the equivalent stress range and the number of stress cycles per truck passage. For variable amplitude stress cycles, the Palmgren-Miner damage law is often used (Byers et al. 1997-a) as $D=\Sigma n_i/N_i$, where $n_i =$ number of stress cycles of stress range i; and N_i is the number of stress cycles to failure in the structural component if the stress range were S_i. From a fracture mechanics approach, fatigue life can be expressed in terms of cycles to failure $N_i = A \cdot S_i^{-m}$ (Kwon and Frangopol 2010). On a basis of equivalent fatigue damage, a revised equivalent stress range, S_w, is used to combine the two parameters for simplifications; namely, the fatigue damage of multiple stress cycles due to each truck passage is considered as the same as that of a single stress cycle of S_w (Zhang and Cai 2011). The fatigue damage from one stress cycle of S_w is $D = A^{-1} \cdot S_w^m$ and equals to the fatigue damage from multiple variable stress ranges $D = A^{-1} \cdot n \cdot S_{re}^m$. For truck passage j, the revised equivalent stress range is:

$$S_w^j = \left(N_c^j\right)^{1/m} \cdot S_{re}^j \tag{3-14}$$

where N_c^j is the number of stress cycles due to the jth truck passage, S_{re}^j is the equivalent stress range of the stress cycles by the jth truck, and m is the material constant that could be assumed as 3.0 for all fatigue categories (Keating and Fisher 1986).

For each truck-passing-bridge analysis, cycle counting methods, such as rainflow counting method, are used to obtain the number of cycles per truck passage. Since the stress range cut-off levels change the number of cycles greatly, a reasonable value is necessary. In the data analysis of stress ranges obtained from field monitoring, 3.45 Mpa (0.5ksi) is a typically used cut-off level for stress ranges to calculate the numbers per truck passage. A similar cut-off level from 3.45 Mpa (0.5ksi) to 33% of the constant amplitude fatigue limit (CAFL) was also suggested by Kwon and Frangopol (2010). Since the contribution of stress ranges less than 3.45 Mpa (0.5ksi) can be neglected, the cut-off level of the stress range of 3.45 Mpa (0.5ksi) is chosen in the present study.

The long undulations in the roadway pavement are assumed as a zero-mean stationary Gaussian random process. The same road roughness coefficient corresponds to randomly generated road profiles and might have varied stress range histories for each road profile. As the output from the dynamic analysis, the stress range history of vehicle passing on the bridge during its life cycle can be simplified as the time history of the equivalent stress range with a reduced length, which could be treated as a random process. Based on previous studies, both the normal and lognormal distribution are acceptable to describe the distribution of the revised equivalent stress range at each combination of road roughness condition and vehicle speed (Zhang and Cai 2011).

3.3.2. Nominal Live Load Stress Range

For load-induced fatigue considerations, each detail should satisfy (AASHTO LRFD 2010):

$$\gamma\left(\Delta f\right) \le \left(\Delta F\right)_n \tag{3-15}$$

where: γ is the load factor, Δf is the live load stress range due to the passage of the fatigue load, and $\left(\Delta F\right)_n$ is the nominal fatigue resistance.

The live load stress range is a random variable. For the convenience of fatigue analysis, a nominal live load stress range, S_{wn}, is defined corresponding to a reliability index β of 3.5, typically used in AASHTO LRFD (2010). In other words, the probability of S_{wn} not being exceeded by the real live load stress ranges corresponds to the reliability index of 3.5. The nominal live load stress range is predicted based on 20 randomly generated road profiles for the given vehicle speed and road roughness coefficient. If the cumulative distribution functions of the live load stress range are defined as F, the nominal live load stress range, S_{wn}, can be calculated as:

$$S_{wn} = F^{-1}\left[\Phi(\beta)\right]$$
(3-16)

where $\Phi(\cdots)$ denotes the standard normal cumulative distribution function.

Different trucks travel on bridges during their service life with a different vehicle speed on a different road condition. The nominal equivalent stress range in the life cycle can be obtained as:

$$S_{wn}^{lc} = \left[\sum_j (p_j)\cdot(S_{wn}^j)^m\right]^{1/m}$$
(3-17)

where p_j means the probability of case j, and here case j is defined as a combination of vehicle type, vehicle speed, road roughness condition and lane numbers.

The reliability based dynamic amplification factor on revised equivalent stress ranges (DAFS) can be defined and obtained as:

$$DAFS^{lc} = \frac{S_{wn}^{lc}}{S_{st}}$$
(3-18)

where S_{st} is the maximum static stress range due to the passage of the live loads without considering the dynamic effects.

In comparison, the dynamic amplification factor (DAF) based on maximum responses is defined as (Paultre et al. 1992):

$$DAF = \frac{R_{dyn}}{R_{sta}} = 1 + IM = \frac{S_{dyn}}{S_{st}}$$
(3-19)

where R_{dyn} is the maximum dynamic response, R_{sta} is the maximum static response, and S_{dyn} is the maximum stress range.

Eq. (3-18), similar to Eq. (3-19) in format, can be conveniently used in fatigue design. For example, when the maximum static stress range S_{st} and DAFS are known, the reliability based nominal live load stress range S_{wn} can be calculated, which includes the fatigue damages from multiple stress range cycles due to each vehicle passage with varied vehicle speeds at various road conditions in the bridge's life cycle. In comparison, Eq. (3-19) is based on a deterministic ratio of the maximum dynamic response and the maximum static response, which could underestimate the actual stress range as discussed earlier.

3.3.3. DAFS for Various Cases

DAFS for the cases with three truck types and six vehicle speeds are listed in Figs. 3-7 to 3-9, respectively.

For the 2-axle trucks, DAFS are no larger than 1.5 for the very good and average road conditions without faulting. When the faulting exists, most of DAFS increase to the range of 1.5 to 6. In the case of 50m/s vehicle speed, DAFS increases to 8.2. Since multiple stress ranges are generated when the vehicle travels along the bridge, the DAFS value may be much larger than the value of DAF that only considers one maximum stress range of the vehicle passing along the bridge.

However, if only one cycle is generated by the vehicle, the DAFS equals to the DAF. For the very poor road condition cases, the relative faulting effect is less than the better road conditions from very good to average and the range of DAFS only shifts from 5.8-7.9 to 4.3-13.1.

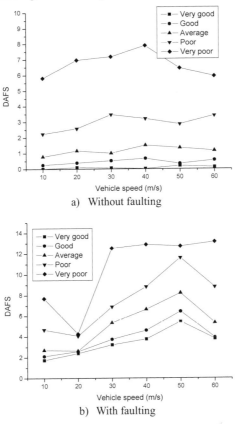

a) Without faulting

b) With faulting

Fig. 3-7 DAFS values for two axle truck

57

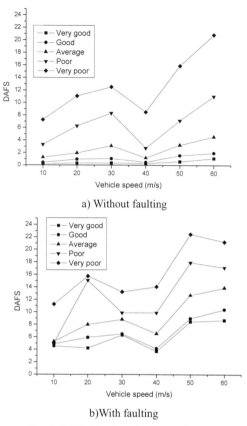

a) Without faulting

b) With faulting

Fig. 3-8 DAFS values for three axle truck

For 3-axle trucks in the cases without faulting, the DAFS values remain small as most of them are less than 2 for the road conditions from very good to average. Large increases of DAFS to the range of 4 to 10 are found if there is faulting. In comparison, the stress range only has a mild increase from 4-20 to 6-22 due to the faulting for the poor and very poor road conditions. The same trends can be found for the 5 axle-trucks. For the very good and good road conditions without faulting, DAFS are less than 2 for most cases. The faulting increases DAFS to the range of 4 to 10. Mild increases could be found for the average and poor cases from range of 3 to 8 to the range of 6 to 12. However, no obvious range difference for DAFS can be found for very poor road conditions due to faulting.

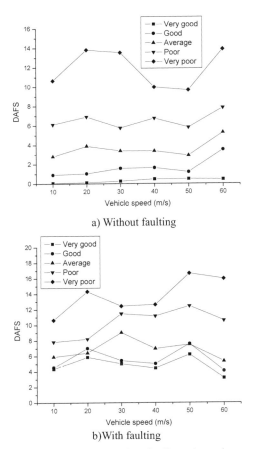

a) Without faulting

b)With faulting

Fig. 3-9 DAFS values for five axle truck

For all the three types of trucks, the faulting induces a large increase of stress range for road conditions from very good to average. When the road conditions deteriorate to poor or very poor, the faulting effects are relatively small and decrease to an ignorable level. The vehicle speed generally increases DAFS but with a limited effect. In some cases, DAFS might decrease with the increase of the vehicle speed. This phenomenon was also reported in the literature (Shi et al. 2008).

Under the normal design condition, calculated fatigue damage equivalent stress range and DAFS is 58 Mpa (8.4 ksi) and 4.0, respectively. It is noteworthy that the length of bridge's life does not affect the DAFS if the percentile of each road roughness condition, vehicle speed distribution and vehicle type in the life cycle remain unchanged.

3.3.4. Parametric Study of DAFS in Life Cycle

DAFS is used in the present study to calculate the dynamic effects from vehicles on bridges' stress range. In the bridge's life cycle, DAFS can be changed by multiple parameters, for instance, faulting, vehicle speed limit and its coefficient of variation, vehicle type, and the annual traffic increase rate. The effects of these parameters are shown in Fig. 3-10.

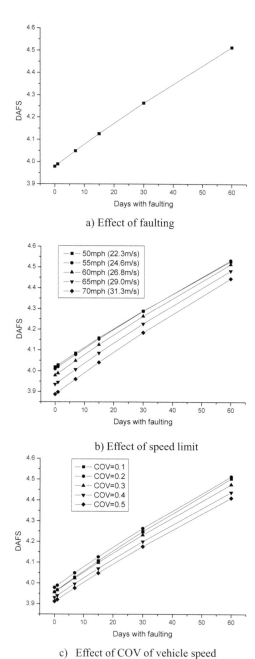

a) Effect of faulting

b) Effect of speed limit

c) Effect of COV of vehicle speed

Fig. 3-10 Parametric study of DAFC in life cycle

(Fig. 3-10 continued)

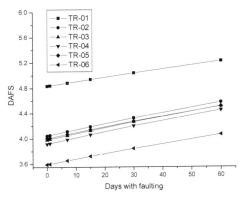

d) Effect of truck distribution

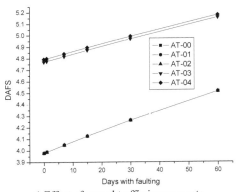

e) Effect of annual traffic increase rate

Twofold road surface condition is used to include the randomly generated road profile from the zero-mean stationary Gaussian random process and the surface discontinuities modeled as faulting. At each year in the bridge's design life, the DAFS value change with the numbers of days with faulting as shown in Fig. 3-10 (a). Generally, the DAFS values increase with the number of days with faulting. When there is no faulting or only one day with faulting, the DAFS value is 4.0. However, when the faulting days increase to one or two months, the DAFS values have a 7% and 13% increase to 4.3 and 4.5, respectively.

Two parameters are used to define the vehicle speed distribution in the present study. One is the vehicle speed limit, and the other is its coefficient of variance. Limited effects from the vehicle speed limit to DAFS are found as shown in Fig. 3-10(b). The DAFS only varies about 0.1 for all the cases with the same days of faulting but with different speed limits ranging from 50mph (22.3 m/s) to 70mph (31.3m/s). The coefficient of variance also has only limited effects on DAFS as shown in Fig. 3-10(c), since the DAFS only varies up to 0.1 for the cases with up to 60 days of faulting.

Under the normal design condition, the average daily truck traffic for the 2, 3 and 5 axle trucks

61

are chosen as 600, 400 and 1000, respectively. Based on the progressive deterioration model for the road roughness as presented in Eq. (3-13), for each road resurface period of 13 years, 7, 2, 2, 1 and 1 years are classified as very good, good, average, poor and very poor road conditions, respectively. Since the 3-axle truck HS20-44 is the design live load for the prototype bridge, the six TR cases are defined according to the ADTT numbers of 3-axle. The numbers of trucks and years of road surface conditions for the TR cases are listed in Table 3-4 (all have a 13 years of resurface period except for TR-01). From Case TR-01 to Case TR-05, the ADTT numbers of 3-axle truck increase from 0 to 800 while the ADTT numbers of 2-axle truck and 5 axle truck decrease from 800 to 400 and 1200 to 800, respectively. In case TR-06, all the 2000 trucks are 3-axle. However, the total ADTT numbers remain unchanged in all the TR cases from TR-01 to TR-06. The DAFSs are not sensitive for the case from TR-02 to TR-05. They are all around 4.0 when there is at most one day of faulting. However, the DAFSs increase greatly in case TR-01, which suggests larger fatigue damage and a shorter fatigue life expectation.

Table 3-4 Number of years for cases with different truck combinations

Case	ADTT for 2,3 & 5 axle trucks	Total	Very good	Good	Average	Poor	Very Poor
TR-01	800, 0, 1200	14	7	2	2	1	2
TR-02	700, 200, 1100	13	7	2	2	1	1
TR-03	600, 400, 1000	13	7	2	2	1	1
TR-04	500, 600, 900	13	7	2	2	1	1
TR-05	400, 800, 800	13	7	2	1	2	1
TR-06	0, 2000, 0	13	7	2	1	1	2

Table 3-5 Number of years for cases with different annual traffic increase rates

Case	α	Total	Very good	Good	Average	Poor	Very Poor
AT-00	0	13	7	2	2	1	1
AT-01	0.01	13	7	2	2	1	1
AT-02	0.02	13	7	2	2	1	1
AT-03	0.03	13	7	2	1	1	2
AT-04	0.05	13	7	1	2	1	2

Under the normal design condition, the annual traffic increase rate is set as zero, which is defined as case AT-00. The cases of AT-01 to AT-04 are defined with an annual traffic increase rate of 1%, 2%, 3% and 5%, respectively. Accordingly, the numbers of years of different road conditions in each resurface period are different as listed in Table 3-5 for each case due to the increase of the fatigue damage from the increased number of trucks. For each road resurface period of 13 years of AT-00, AT-01, and At-02, 7, 2, 2, 1 and 1 years are classified as very good, good, average, poor and very poor road conditions, respectively. There are nearly no differences of DAFS for the cases AT-00, AT-01, and AT-02 since they have the same distribution of road surface condition distribution. However, for the cases AT-03 and AT-04, 7, 2, 1, 1, 2 years and 7, 1, 2, 1, 2 years are classified as very good, good, average, poor and very poor road conditions, respectively. An increase of DAFS from 4.0 to 4.8 is found for the cases AT-03 and AT04, which also suggests a shorter fatigue life expectation.

3.4 Fatigue Life Estimation

3.4.1. Various Approaches for Fatigue Life Estimation

Fatigue life of the bridge can be obtained through either deterministic or probabilistic approach. To demonstrate the proposed methodology, six different approaches are used in the present study to obtain the fatigue life at a given reliability index and are compared with each other. The DT-DAF corresponds to the AASHTO LRFD (2010) deterministic fatigue analysis methodology. The DT-DAFS is the same as DT-DAF except using the proposed DAFS to replace the DAF. In the PB-DAFS and PB-DAF approaches, a probabilistic fatigue analysis is conducted based on a limit state function, using the deterministic DAFS and DAF, respectively. For the purpose of comparison, in the PB-SWE and PB-SWM approaches, instead of using the developed deterministic DAFS and DAF, the equivalent stress ranges are treated as random variables in the limit state function. The PB-SWE approach includes all the stress ranges in one vehicle-passing-bridge analysis, while only the maximum stress range is included in the PB-SWM approach.

3.4.2. Deterministic Approach

In the AASHTO LRFD bridge design specifications, the deterministic approach is used. The design criteria of load-induced fatigue are presented in Eq. (3-15). If the DAFS is used, the live load stress range Δf is taken as:

$$\Delta f = S_{wn} = DAFS \cdot S_{st}^{HS} \tag{3-20}$$

The nominal fatigue resistance is taken as (AASHTO LRFD 2010):

$$(\Delta F)_n = \left(\frac{A}{N}\right)^{\frac{1}{3}} \ge \frac{1}{2}(\Delta F)_{TH} \tag{3-21}$$

in which :

$$N = (365)(Years)n(ADTT)_{SL} \tag{3-22}$$

where A is the detail constant taken from Table 6.6.1.2.5-1 in AASHTO LRFD bridge design specifications (AASHTO 2010); n is the number of stress range cycles per truck passage taken from Table 6.6.1.2.5-2. Since the revised equivalent stress ranges have enclosed both of the stress cycles and the stress ranges, n is one (1) in the current deterministic approach; $(ADTT)_{SL}$ is single-lane ADTT and $(\Delta F)_{TH}$ is the constant-amplitude fatigue threshold taken from Table 6.6.1.2.5-3.

If the traditional dynamic amplification factor prescribed in AASHTO LRFD is used (DT-DAF approach), the live load stress range Δf equals to DAF × 14.5 Mpa (2.1 ksi) = 41 Mpa (5.9 ksi), which is less than a half of the threshold for the Category B. Therefore, the fatigue life of the bridge detail is infinite. In comparison, the fatigue life is calculated as 94 years for the bridge under normal design condition when the DAFS is used to obtain the live load stress range (DT-DAFS approach).

3.4.3. Probabilistic Approach

In the probabilistic approach, a limit state function (LSF) needs to be defined first in order to ensure a target fatigue reliability (Nyman and Moses 1985):

$$g(X) = D_f - D(t) \tag{3-23}$$

where D_f is the damage to cause failure and is treated as a random variable with a mean value of 1; $D(t)$ is the accumulated damage at time t; and g is a failure function such that $g<0$ implies a fatigue failure. The overall fatigue damages are a summation of damages done by the trucks under all vehicle speed ranges, lane numbers and road roughness conditions. The accumulated damage $D(t)$ is (Zhang and Cai 2011):

63

$$D(t) = \sum_j \frac{n_j}{N_j} = \sum_j \frac{n_{truck}^j \cdot N_c}{A \cdot \left(S_{re}^j\right)^{-m}} = n_{tr} \cdot A^{-1} \cdot \sum_j \left(p_j\right) \cdot \left(S_w^j\right)^m \tag{3-24}$$

Therefore, the limit state function is:

$$g(X) = D_f - n_{tr} \cdot A^{-1} \cdot \sum_j \left(p_j\right) \cdot \left(S_w^j\right)^m \tag{3-25}$$

Table 3-6 Summary of LSF parameters

Par.	Mean	COV	Distribution	Description
D_f	1.0	0.15	Lognormal	Damage to cause failure
ADTT	2000		Deterministic	ADTT in fatigue life
N_c	Calculated			Number of cycles per truck passage
t	75		Deterministic	Total fatigue life in years
A	7.83×10^{10}	0.34	Lognormal	Detail constant
m	3.0		Deterministic	Slope constant
S_w	Calculated	Calculated	Lognormal	Revised equivalent stress range
V	50.0mph (22.4m/s)	0.2	normal	Vehicle speed
$DAFC$	Calculated	Calculated	Deterministic	Reliability based DAF on stress ranges

Based on the information from the literature, all the related random variables for predicting fatigue reliabilities are listed in Table 3-6, including their distribution types, mean values, coefficient of variations (COVs) and descriptions (Zhang and Cai 2011). As a result, the fatigue reliability index can be obtained based on Eq. (3-25) (PB-SWE approach). If only the maximum stress range is used to calculate S_w in Eq. (3-25), the corresponding results of PB-SWM approach can be obtained.

Replacing S_w with S_{st}^{HS} and DAFS using Eqs. (3-17) and (3-18), the limit state function can be changed and rearranged for simplicity as:

$$g(X) = D_f \cdot A - n_{tr} \cdot \left(S_{st}^{HS} \cdot DAFS^{lc}\right)^m \tag{3-26}$$

After introducing the DAFS, only two random variables are left in Eq. (3-26), namely, D_f and A. Both of them follow lognormal distribution and the fatigue reliability index can be easily obtained (PB-DAFS approach). Similarly, by replacing DAFS with DAF in Eq. (3-26), a PB-DAF analysis can be carried out.

In the present study, the target reliability index β is chosen as 3.5, a value typically used in AASHTO LRFD (2010). For the chosen target reliability index, fatigue life 66 years and 60 years for PB-DAFS and PB-SWE, respectively, can be obtained via the probabilistic approach under normal design condition. Similarly, fatigue life can also be obtained if using the maximum-value-based traditional DAF or treating the maximum value of stress range as random variables in the LSF. The calculated fatigue life is 194 years and 225 years for PB-DAF and PB-SWM, respectively. Further discussions of the 6 approaches are given next.

3.4.4. Comparisons of Different Approaches

In order to appreciate the difference of the proposed DAFS and traditional DAF, the calculated fatigue lives from the six approaches are compared with each other for the same target reliability index β=3.5. The DAFS and DAF, and the corresponding fatigue lives with varied faulting days in each year are shown in Fig. 3-11.

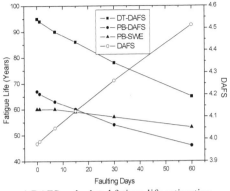

a) DAFS and related fatigue life estimation

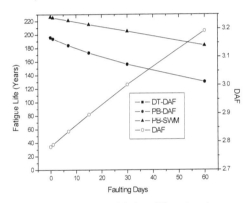

b) DAF and related fatigue life estimation

Fig. 3-11 Fatigue life estimation

While DAF only reflects the largest stress amplitude during one vehicle passing on the bridge, DAFS includes the fatigue damages from multiple stress range cycles due to each vehicle passage. The DAF is less than the DAFS and leads to an overestimation of fatigue life. As a result, the DAF as shown in Fig. 3-11(a) is about 30% less than the DAFS as shown in Fig. 3-11(b). Correspondingly, the fatigue lives using the DAF are overestimated to a scale of 3 to 4 compared with that using the DAFS. The deterministic approaches DT-DAFS and DT-DAF predict a longer fatigue life than the probabilistic approaches PB-DAFS & PB-SWE and PB-DAF & PB-SWM, respectively. The fatigue lives from the approach DT-DAFS is increased by 20% to 60% compared with the results from the approaches PB-DAFS and PB-SWE as shown in Fig. 3-11(a), while the predicted fatigue lives are infinite for the deterministic approach DT-DAF.

As mentioned above, the numbers of faulting days have a large effect on the DAFS and thus on the fatigue life estimation. As shown in Fig. 3-11(a), if the faulting day is less than a half month, no fatigue life is lost based on the   PB-SWE approach and 4 years of fatigue life are lost based on PB-DAFS. However, when the faulting days increase to one month, the fatigue lives decrease from

60 years and 67 years to 57 years and 54 years, respectively. Based on these results, major road damage, such as a path hole, should be repaired within a half month period.

Based on the PB-SWE approach, the fatigue life of all the 84 cases with varied faulting days, speed limits, COV of vehicle speeds, and truck distributions are calculated and shown in Fig. 3-12. Correspondingly, the fatigue lives obtained through the approaches of PB-DAFS and DT-DAFS are plotted in the figure, as well. The fatigue lives decrease with the increase of the DAFS for the approaches of PB-DAFS and DT-DAFS. All the data sets obtained from PB-SWE approach are in-between the results from the PB-DAFS and DT-DAFS approaches. The large differences between the two methods (DT-DAFS and PB-DAFS) originate from the load factor γ. In AASHTO LRFD (2010), the load factor γ is to reflect the load level found to be representative of the truck population with respect to a large number of return cycles of stresses and to their cumulative effects. Since the truck distribution and varied stress cycles for different trucks have been considered in the present study, the load factor $\gamma = 0.75$ is not necessary in the DT-DAFS approach. The recalculated results for $\gamma = 1.0$ are labeled as DT-DAFS-COF in the figure and they are much closer to those of PB-DAFS.

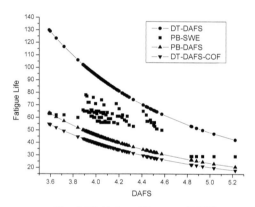

Fig. 3-12 Fatigue life versus DAFS

3.5 Conclusions

In the present study, a reliability based dynamic amplification factor on revised equivalent stress ranges (DAFS) for fatigue design is proposed to include the fatigue damages from multiple stress range cycles due to each vehicle passage at varied vehicle speeds under various road conditions in the bridge's life cycle. The effects of the long-term deck deterioration and various vehicle parameters, such as vehicle speeds and types, can be included in DAFS, as well. A numerical simulation toward solving a coupled vehicle-bridge system including a 3-D suspension vehicle model and a 3-D dynamic bridge model is used to obtain the revised equivalent stress range. Parametric studies of DAFS are carried out to find the effects from multiple variables in the bridge's life cycle, for instance, the faulting days in each year, vehicle speed limit and its coefficient of variance, vehicle type distribution, and annual traffic increase rate. The calculated fatigue lives from the six different approaches, namely, DT-DAFS, PB-DAFS, PB-SWE, DT-DAF, PB-DAF, and PB-SWM, are compared with each other to acquire a reasonable fatigue life estimation to preserve both simplicity and accuracy. From the present study, the following conclusions are drawn:

1. DAFS is an effective measure of dynamic stress cycles that can include the effects from random variables in the vehicle-bridge dynamic system. Under the same target reliability level, a larger DAFS value corresponds to shorter fatigue lives.

2. Faulting in the road surface increases the DAFS values and decreases the fatigue life. It has limited influence when the damages are repaired within 15 days for most cases in the present study.

3. DAFS is sensitive to the road roughness deterioration rate in the bridge's life cycle. The effects of vehicle type, annual traffic increase rate, and some other parameters are reflected by the DAFS via the change of road roughness conditions in each road resurface period.

4. Since DAF only reflects the largest stress amplitude while DAFS includes the fatigue damages from multiple stress range cycles due to each vehicle passage, DAF is less than the DAFS and leads to an overestimation of fatigue life.

The present study has demonstrated the methodology through a prototype bridge. Based on the obtained DAFS and the static effect of the wheel loads, the dynamic stress ranges and fatigue life of the bridge can be easily obtained. Both the accuracy and simplicity for bridge fatigue design can be preserved. However, the process of obtaining DAFS is based on a quite complicated theoretical approach. More numerical simulations and sensitivity studies are needed to recommend design DAFS values for the small and medium bridges with varied structural dynamic properties in the future study. After evaluation of the current condition of the structure, certain actions might be taken. Multiple DAFS values can be defined in bridge's life cycle based on the past traffic and road condition records and the predictions of future traffic and road deterioration rate. Therefore, reasonable maintenance strategies can be implemented to ensure the safety and reliability of existing bridges.

3.6 References

Agarwal, A. C., and Billing, J. R. (1990) "Dynamic testing of the St. Vincent Street Bridge." *Proceedings of the Annual Conference of the Canadian Society for Civil Engineering*, Hamilton, Ont., 163-181.

ASCE Committe on Fatigue and Fracture Reliability of the Committee on structural safety and reliability of the structural division (1982). "Fatigue Reliability." *Journal of the Structural Division - Proceedings of the American Society of Civil Engineers*, ASCE Committee on Fatigue and Fracture Reliability, 108(ST1), 3-88.

Ashebo, D. B., Chan, T. H. T., and Yu, L. (2007). "Evaluation of dynamic loads on a skew box girder continuous bridge Part II: Parametric study and dynamic load factor." *Engineering Structures*, 29(6), 1064-1073.

American Association of State Highway and Transportation Officials (AASHTO). (2010). LRFD bridge design specifications, Washington, DC.

Billing, J.R. (1984). "Dynamic loading and testing of bridges in Ontario." *Canadian Journal of Civil Engineering*, 11, 833-843.

Byers, W. G., Marley, M. J., Mohammadi, J., Nielsen, R. J., and Sarkani, S. (1997a). "Fatigue Reliability Reassessment Procedures: State-of-the-Art Paper." *Journal of Structural Engineering*, 123(3), 271-276.

Byers, W. G., Marley, M. J., Mohammadi, J., Nielsen, R. J., and Sarkani, S. (1997b). "Fatigue Reliability Reassessment Applications: State-of-the-Art Paper." *Journal of Structural Engineering*, 123(3), 277-285.

Cai, C. S., and Chen, S. R. (2004). "Framework of vehicle-bridge-wind dynamic analysis." *Journal of Wind Engineering and Industrial Aerodynamics*, 92(7-8), 579-607.

Cai, C. S., Shi, X. M., Araujo, M., and Chen, S. R. (2007). "Effect of approach span condition on vehicle-induced dynamic response of slab-on-girder road bridges." *Engineering Structures*, 29(12), 3210-3226.

Cai, C.S., Shi, X.M., Voyiadjis, G.Z. and Zhang, Z.J. (2005). "Structural performance of bridge approach slab under given embankment settlement." *Journal of Bridge Engineering*, 10(4), 482-489.

Cai, C.S., Zhang, W., Xia, M. and Deng, L. (2011). "A reliability based simulation, monitoring and code calibration of vehicle effects on existing bridge performance" Proceeding, The 5th Cross-strait Conference on Structural and Geotechnical Engineering, July 13-15, Hong Kong, China.

Cebon, D. (1999). Handbook of vehicle-road interaction, 2nd Ed., Swets & Zeitlinger B.V., Lisse, the Netherlands.

Cheung, M. S., and Li, W. C. (2003). "Probabilistic fatigue and fracture analyses of steel bridges." *Structural Safety*, 25(3), 245-262.

Chung, H. (2004). "Fatigue reliability and optimal inspection strategies for steel bridges," Ph.D Dissertation, The University of Texas at Austin.

Chung, H.-Y., Manuel, L., and Frank, K. H. (2006). "Optimal Inspection Scheduling of Steel Bridges Using Nondestructive Testing Techniques." *Journal of Bridge Engineering*, 11(3), 305-319.

Deng, L., and Cai, C. S. (2010). "Development of dynamic impact factor for performance evaluation of existing multi-girder concrete bridges." *Engineering Structures*, 32(1), 21-31.

Deng, L., Cai, C. S. and Barbato, M. (2011). "Reliability-based dynamic load allowance for capacity rating of prestressed concrete girder bridges." *Journal of Bridge Engineering*, Accepted.

Ding, L., Hao, H., and Zhu, X. (2009). "Evaluation of dynamic vehicle axle loads on bridges with different surface conditions." *Journal of Sound and Vibration*, 323(3-5), 826-848.

Dodds, C. J., and Robson, J. D. (1973). "The Description of Road Surface Roughness." *Journal of Sound and Vibration*, 31(2), 175-183.

Donnell, E. T., Hines, S. C., Mahoney, K. M., Porter, R. J., and McGee, H. (2009). "Speed Concepts: Informational Guide." U.S. Department of Transportation and Federal Highway Administration. Publication No. FHWA-SA-10-001.

Green, M. F. (1990). "The dynamic response of short-span highway bridges to heavy vehicle loads," Ph.D. Dissertation, University of Cambridge, Cambridge, UK.

Green, M. F. (1993). "Discussion: Bridge dynamics and dynamic amplification factors - a review of analytical and experimental findings." *Canadian Journal of Civil Engineering*, 20(5), 876-877.

Green, M.F., Cebon, D. and Cole, D.J. (1997). "Effects of heavy vehicle suspension design on the dynamics of highway bridges." *Journal of Structural Engineering*, 121 (2), 272-282.

Guo, W. H., and Xu, Y. L. (2001). "Fully computerized approach to study cable-stayed bridge-vehicle interaction." *Journal of Sound and Vibration*, 248(4), 745-761.

Honda, H., Kajikawa, Y., and Kobori, T. (1982). "Spectra of Road Surface Roughness on Bridges." *Journal of the Structural Division*, 108(ST-9), 1956-1966.

Hwang, E.S., and Nowark, A.S. (1991). "Simulation of dynamic load for bridges." *Journal of Structural Engineering*, 117 (5), 1413-1434.

International Standard Organization. (1995). "Mechanical vibration - Road surface profiles - Reporting of measured data." Geneva.

Keating, P. B., and Fisher, J. W. (1986). "Evaluation of Fatigue Tests and Design Criteria on Welded Details." NCHRP Report 286, Transportation Research Board, Washington, D.C.

Kwon, K., and Frangopol, D. M. (2010). "Bridge fatigue reliability assessment using probability density functions of equivalent stress range based on field monitoring data." *International Journal of Fatigue*, 32(8), 1221-1232.

Laman, J. A., and Nowak, A. S. (1996). "Fatigue-Load Models for Girder Bridges." *Journal of Structural Engineering*, 122(7), 726-733.

Nassif, H. H., Liu, M., and Ertekin, O. (2003). "Model Validation for Bridge-Road-Vehicle Dynamic Interaction System." *Journal of Bridge Engineering*, 8(2), 112-119.

Miller, J.S. and Bellinger, W.Y. "Distress Identification Manual for the Long-Term Pavement Performance Program (Fourth Revised Edition)" Federal Highway Administration, U.S. Department of Transportation, Publication NO. FHWA-RD-03-031, Jun 2003.

Nyman, W. E., and Moses, F. (1985). "Calibration of Bridge Fatigue Design Model." *Journal of Structural Engineering*, 111(6), 1251-1266.

O'Connor, C, and Pritchard, R.W. (1985) "Impact studies on small composite girder bridges." Journal of Structural Engineering, 111, 641-653.

Park, Y. S., Shin, D. K., and Chung, T. J. (2005). "Influence of road surface roughness on dynamic impact factor of bridge by full-scale dynamic testing." *Canadian Journal of Civil Engineering*, 32(5), 825-829.

Paultre, P., Chaallal, O., and Proulx, J. (1992). "Bridge dynamics and dynamic amplification factors - a review of analytical and experimental findings." *Canadian Journal of Civil Engineering*, 19(2), 260-278.

Pipinato, A., Pellegrino, C., and Modena, C. (2011). "Fatigue assessment of highway steel bridges in presence of seismic loading." *Engineering Structures*, 33(1), 202-209.

Pourzeynali, S., and Datta, T. K. (2005). "Reliability Analysis of Suspension Bridges against

Fatigue Failure from the Gusting of Wind." *Journal of Bridge Engineering*, 10(3), 262-271.

Shi, X., Cai, C. S., and Chen, S. (2008). "Vehicle Induced Dynamic Behavior of Short-Span Slab Bridges Considering Effect of Approach Slab Condition." *Journal of Bridge Engineering*, 13(1), 83-92.

TxDOT. (2006). "Procedures for Establishing Speed Zones." Texas Department of Transportation.

Wang, T. L., and Huang, D. (1992). "Computer modeling analysis in bridge evaluation." Florida Department of Transportation, Tallahassee, FL.

Wang, T. L, and Liu, C.H. (2000). "Influence of Heavy Trucks on Highway Bridges," Rep. No. FL/DOT/RMC/6672-379, Florida Department of Transportation, Tallahassee, FL.

Zhang, W., C.S. Cai. (2011). "Fatigue Reliability Assessment for Existing Bridges Considering Vehicle and Road Surface Conditions", Journal of Bridge Engineering, doi:10.1061/ (ASCE) BE. 1943-5592.0000272

CHAPTER 4   PROGRESSIVE FATIGUE RELIABILITY ASSESSMENT OF EXISTING BRIDGES BASED ON A NONLINEAR CONTINUOUS FATIGUE DAMAGE MODEL

4.1 Introduction

During the life cycle of a bridge, the varying dynamic loading from vehicles on the deteriorated road surfaces can lead to fatigue damage accumulations in structure details. Such damages might develop into micro cracks and lead to serious fatigue failures for bridge components or a whole structure failure, for instance, the collapse and failure of the Point Pleasant Bridge in West Virginia (1967) and Yellow Mill Pond Bridge in Connecticut (1976). A life prediction and reliability evaluation is challenging despite the extensive progress on the modeling of vehicle-bridge dynamic interactions and fatigue damage accumulation rules (Guo and Xu 2001, Cai and Chen 2004, Liu and Mahadeven 2007, Yao et al. 1986).

Under constant amplitude loadings, the relationship between the fatigue life and the stress level can be achieved via coupon testing, and S-N curves are obtained from the tests. However, for most bridge details in practice, the stresses generated by repeated dynamic loading have varying amplitude ranges. Compared with the fatigue issues under constant amplitude loadings, it is more difficult to model the fatigue problems correctly under varying amplitude loadings. A more accurate fatigue damage accumulation rule is required. The linear damage rule (LDR) proposed by Miner (1945) is easy and frequently used. However, it may not be sufficient to describe the physics of fatigue damage accumulations (Fatemi and Yang 1998), and a large scatter in the fatigue life prediction can be found (Shimokawa and Tanaka 1980, Kawai and Hachinohe 2002, Yao et al. 1986). During the most part of bridges' fatigue lives, the structure materials are in a linear range, and micro cracks have not developed into macroscopic cracks. After the initial crack propagation stage, the fatigue damage accumulation can be predicted through fracture mechanics analyses. However, the fatigue life assessment of existing bridges is related to a sequence of progressive fatigue damages with only the initiations of micro cracks. Nonlinear cumulative fatigue damage theories were developed to model the fatigue damage accumulation in this stage (Arnold and Kruch 1994, Chabache and Lesne 1988). These theories are based either on separation of fatigue life into two periods (initiation and propagations) or on remaining life and continuous damage concepts. Therefore, the nonlinear continuous fatigue damage model is more appropriate for the fatigue analysis during a large fraction of bridges' life cycles. However, there are no systematic approaches on progressive fatigue reliability assessments available to include multiple dynamic loads and progressively deteriorated road surface conditions.

In the present study, a progressive fatigue reliability assessment approach is proposed based on the nonlinear continuous fatigue damage model. The paper is organized as the following three main sections. In the first section, the process of generating stress range histories is detailed. After introducing the vehicle-bridge dynamic system, the principles for generating progressive deteriorated road profiles are introduced. In the second section, linear and nonlinear continuous fatigue damage rules are introduced. Based on these fatigue damage rules, a progressive fatigue reliability assessment approach is proposed. The fatigue life and fatigue reliability index can be obtained including multiple random variables of the vehicle-bridge dynamic system in a bridge's life cycle. In the third section, a numerical example on the fatigue reliability assessment is presented, and the effect of the fatigue damage rules, surface discontinuities, and vehicle speeds on the fatigue life estimation are discussed.

4.2 Generating Stress Range History

4.2.1. Vehicle-Bridge Dynamic System

From modeling the vehicle loads as a constant moving force (Timoshenko et al. 1974) or moving mass (Blejwas et al. 1979) to using a full vehicle-bridge coupled model (Guo and Xu 2001), the structural analysis of bridges has been extended to the dynamic analysis of a structural system under multiple dynamic loads to count their coupled effects (Cai and Chen 2004, Chen et al. 2011). In the coupled analysis, the interactions between the bridge and vehicles are modeled as coupling forces between the tires and the road surface. As such, the coupling forces were proven to be significantly affected by the vehicle speed and road roughness conditions and resulted in significant effects on the dynamic responses of short span bridges (Deng and Cai 2010, Shi et al. 2008, Zhang and Cai 2011). In the present study, the vehicle is modeled as a combination of several rigid bodies connected by several axle mass blocks, springs, and damping devices (Cai and Chen 2004), and the tires and suspension systems are idealized as linear elastic spring elements and dashpots.

The equations of motions for the vehicle and the bridge are expressed as:

$$[M_v]\{\ddot{d}_v\}+[C_v]\{\dot{d}_v\}+[K_v]\{d_v\}=\{F_v^G\}+\{F_c\} \tag{4-1}$$

$$[M_b]\{\ddot{d}_b\}+[C_b]\{\dot{d}_b\}+[K_b]\{d_b\}=\{F_b\} \tag{4-2}$$

where $\{d\}$ are displacement vectors, $[M]$ are the mass matrices, $[C]$ are the damping matrices and $[K]$ are the stiffness matrices, where subscript v is for vehicle and b for bridge; $\{F_b\}$ is the wheel-bridge contact forces on the bridge, $\{F_v^G\}$ is the self-weight of vehicle, and $\{F_c\}$ is the vector of wheel-road contact forces acting on the vehicle. The two equations are coupled through the contact condition, i.e., the interaction forces $\{F_c\}$ and $\{F_b\}$, which are action and reaction forces existing at the contact points of the two systems and can be stated as a function of deformation of the vehicle's lower spring:

$$\{F_b\}=-\{F_c\}=[K_l]\{\Delta_l\}+[C_l]\{\dot{\Delta}_l\} \tag{4-3}$$

where $[K_l]$ and $[C_l]$ are the coefficients of the vehicle's lower spring and damper, and Δ_l is the deformation of the lower springs of the vehicle. The relationship among the vehicle-axle-suspension displacement Z_a, displacement of bridge at wheel-road contact points Z_b, deformation of lower springs of vehicle Δ_l, and road surface profile $r(x)$ are derived as:

$$Z_a = Z_b + r(x) + \Delta_l \tag{4-4}$$

$$\dot{Z}_a = \dot{Z}_b + \dot{r}(x) + \dot{\Delta}_l \tag{4-5}$$

where $\dot{r}(x)=(dr(x)/dx)\cdot(dx/dt)=(dr(x)/dx)\cdot V(t)$ and $V(t)$ is the vehicle velocity.

Therefore, the contact forces $\{F_b\}$ and $\{F_c\}$ between the vehicle and the bridge are derived as:

$$\{F_b\}=-\{F_c\}=[K_l]\{Z_a-Z_b-r(x)\}+[C_l]\{\dot{Z}_a-\dot{Z}_b-\dot{r}(x)\} \tag{4-6}$$

After transforming the contact forces to equivalent nodal forces and substituting them into Eqs. (4-1) and (4-2), the final equations of motion for the coupled system are as follows (Shi et al. 2008):

$$\begin{bmatrix} M_b & \\ & M_v \end{bmatrix} \begin{Bmatrix} \ddot{d}_b \\ \ddot{d}_v \end{Bmatrix} + \begin{bmatrix} C_b + C_{bb} & C_{bv} \\ C_{vb} & C_v \end{bmatrix} \begin{Bmatrix} \dot{d}_b \\ \dot{d}_v \end{Bmatrix} + \begin{bmatrix} K_b + K_{bb} & K_{bv} \\ K_{vb} & K_v \end{bmatrix} \begin{Bmatrix} d_b \\ d_v \end{Bmatrix} = \begin{Bmatrix} F_{br} \\ F_{vr} + F_v^G \end{Bmatrix} \qquad (4\text{-}7)$$

The additional terms C_{bb}, C_{bv}, C_{vb}, K_{bb}, K_{bv}, K_{vb}, F_{br} and F_{vr} in Eq. (4-7) are due to the expansion of the contact force in comparison with Eqs. (4-1) and (4-2). When the vehicle is moving across the bridge, the bridge-vehicle contact points change with the vehicle position and the road roughness at the contact point. As a large number of degrees of freedom (DOF) are involved, the mode superposition technique is used to simplify the modeling procedure based on the obtained bridge mode shape and the corresponding natural circular frequencies.

After obtaining the bridge dynamic response $\{d_b\}$, the stress vector can be obtained by:

$$[S] = [E][B]\{d_b\} \qquad (4\text{-}8)$$

where $[E]$ is the stress-strain relationship matrix and is assumed to be constant over the element, and $[B]$ is the strain-displacement relationship matrix assembled with x, y and z derivatives of the element shape functions.

4.2.2. Progressive Deteriorated Road Profile

In the current AASHTO LRFD specifications (AASHTO 2010), the dynamic effects due to moving vehicles are attributed to two sources, namely, the hammering effect due to the vehicle riding surface discontinuities, such as deck joints, cracks, potholes and delaminations, and dynamic response due to long undulations in the roadway pavement. The long undulations in the roadway pavement can be assumed as a zero-mean stationary Gaussian random process, and it can be generated through an inverse Fourier transformation (Wang and Huang 1992):

$$r(x) = \sum_{k=1}^{N} \sqrt{2\phi(n_k)\Delta n} \cos(2\pi n_k x + \theta_k) \qquad (4\text{-}9)$$

where θ_k is the random phase angle uniformly distributed from 0 to 2π; $\phi()$ is the power spectral density (PSD) function (m³/cycle) for the road surface elevation; n_k is the wave number (cycle/m). The PSD functions for road surface roughness were developed by Dodds and Robson (1973), and the PSD function was simplified by Wang and Huang (1992) as:

$$\phi(n) = \phi(n_0)(\frac{n}{n_0})^{-2} \qquad (4\text{-}10)$$

where $\phi(n)$ is the PSD function (m³/cycle) for the road surface elevation; n is the spatial frequency (cycle/m); n_0 is the discontinuity frequency of $1/2\pi$ (cycle/m); and $\phi(n_0)$ is the road roughness coefficient (m³/cycle) whose value is chosen depending on the road condition.

In order to include the progressive pavement damages due to traffic loads and environmental corrosions, a progressive road roughness deterioration model for the bridge deck surface is used (Zhang and Cai, 2011):

$$\phi_t(n_0) = 6.1972 \times 10^{-9} \times \exp\left\{ \left[8.39 \times 10^{-6} \phi_0 e^{\eta t} + 263(1 + SNC)^{-5} \left(CESAL \right)_t \right] / 0.42808 \right\} + 2 \times 10^{-6} \qquad (4\text{-}11)$$

where ϕ_t is the road roughness coefficient at time t; ϕ_0 is the initial road roughness coefficient directly after completing the construction and before opening to traffic; t is the time in years; η is the environmental coefficient varying from 0.01 to 0.7 depending upon the dry or wet, freezing or non-freezing conditions; SNC is the structural number modified by sub grade strength; and $(CESAL)_t$

is the estimated number of traffic in terms of AASHTO 18-kip cumulative equivalent single axle load at time t in millions.

For the surface discontinuities that cause hammer effects, these irregularities, such as the uneven joints, the potholes, and faulting (bumps), have a significant influence on bridge dynamic response and should be isolated and treated separately from such pseudo-random road surface profiles according to ISO (ISO 8606, 1995) and Cebon (1999). The local unevenness of expansion joints at the approach slab ends was found to significantly increase the dynamic response of short span bridges. The discontinuities can be modeled with a step up or down for the faulting between the approach slab and pavement and between the bridge deck and approach slab (Green et al. 1997, Shi et al. 2008). According to US Federal Highway Administration (Miller and Bellinger 2003), low, moderate and high severity potholes in the pavement are defined as 0.025m, 0.025-0.05m and more than 0.05m deep. In the present study, faulting of 0.038m is used to model the surface discontinuities. The discontinuities are assumed to be located at the entrance of the bridge. However, the vehicle is assumed to be traveling before entering the bridge. In the present study, the faulting values are assumed at both ends of the approach slab with the same values, and the approach slab deflection and slope change in the approach slab was also used in the road profile to describe vehicle riding surface discontinuities (Shi et al. 2008, Cai et al. 2007). Therefore, a twofold road surface condition is used in the vehicle-bridge dynamic analysis to include both the local defects and long undulations of road profiles for dynamic effects due to moving vehicles. Since most of the major road damages are expected to finish in one day, the default day with surface discontinuities in each year is assumed as one.

4.2.3. Revised Equivalent Stress Range

Since each truck passage might induce multiple stress cycles, two correlated parameters are essential to calculate the fatigue damages due to each truck passage, i.e. the equivalent stress range and the number of stress cycles caused per truck passage. On a basis of equivalent fatigue damage, a revised equivalent stress range, S_w, is used to combine the two parameters for simplifications; namely, the fatigue damage of multiple stress cycles due to each truck passage is considered as the same as that of a single stress cycle of S_w (Zhang and Cai 2011). For truck passage j, the revised equivalent stress range is:

$$S_w^j = \left(N_c^j \right)^{1/m} \cdot S_{re}^j \tag{4-12}$$

where N_c^j is the number of stress cycles due to the jth truck passage, S_{re}^j is the equivalent stress range of the stress cycles by the jth truck, and m is the material constant that can be assumed as 3.0 for all fatigue categories (Keating and Fisher 1986).

The long undulations in the roadway pavement are assumed as a zero-mean stationary Gaussian random process. A same road roughness coefficient may correspond to different randomly generated road profiles and, consequently, might result in different stress range histories for each road profile. As the output from the dynamic analysis, the stress range history of vehicles passing on the bridge during its life cycle can be simplified as the time history of the equivalent stress range, which could be treated as a random process. Based on previous studies, both the normal and lognormal distribution are acceptable to describe the distribution of revised equivalent stress ranges at each combination of road roughness conditions and vehicle speeds (Zhang and Cai 2011).

4.3 Fatigue Reliability Assessment

4.3.1. Linear and Non-linear Damage Rule

Fatigue, due to an accumulation of damage, is one of the main forms of deterioration for structures and can be a typical failure mode. Due to the progressive deteriorations and accumulated fatigue damages of structures under dynamic loads, such as vehicles, it is essential to ensure the structure's safety. Among all the fatigue damage accumulation rules, the linear damage accumulation rule (LDR), also known as Miner's rule, is the most commonly used for fatigue damage accumulation of variable loadings (Miner 1945):

$$D(t) = \sum_i \frac{n_i}{N_i} = \frac{n_{tc}}{N} \qquad (4\text{-}13)$$

where $D(t)$ is the accumulated fatigue damage at time t, n_i is number of observations in the predefined stress-range bin S_{ri}, N_i is the number of cycles to failure corresponding to the predefined stress-range bin, n_{tc} is the total number of stress cycles, and N is the number of cycles to failure under an equivalent constant amplitude loading (Kwon and Frangopol 2010):

$$N = A \cdot S_{re}^{-m} \qquad (4\text{-}14)$$

where S_{re} is the equivalent stress range and A is the detail constant that is typically defined in design codes, such as Table 6.6.1.2.5-1 in AASHTO (2010).

Due to its simplicity, the LDR is most widely and frequently used. However, it has been shown that LDR produces a large scatter in the fatigue life prediction of both metal and composites (Shimokawa and Tanaka 1980, Kawai and Hachinohe 2002). In addition, the load level dependence of fatigue damage cannot be explained by the LDR model (Halford 1997). LDR cannot explain that the damage accumulations D obtained in the experiments is larger than 1 for low-high load sequences and smaller than 1 for high-low sequences (Fatemi and Yang 1998).

In order to improve the accuracy of LDR, non-linear cumulative fatigue damage rules had been developed based either on the separation of fatigue life into two periods (initiation and propagation) on the progressive decrease of fatigue limit or on remaining life and continuous damage concepts (Marco and Starkey 1954, Manson and Halford 1981, Chaboche and Lesne 1988). The nonlinear accumulation function proposed by Marco and Starkey (1954) is expressed as:

$$D = \sum_{i=1}^{k} \left(\frac{n_j}{N_j} \right)^{C_i} \qquad (4\text{-}15)$$

where C_i is a material parameter related to i^{th} loading level. The damage curve approach proposed by Manson and Halford (1981) has a similar formula. Based on such a model, both of the load-level dependence and load-sequence dependence effects of the fatigue damage accumulations can be reflected. In order to save the calculation cost, double linear functions were used to approximate the nonlinear function, and a linear damage accumulation rule is applied (Halford 1997). Such simplifications can be easily used in two-block loading problems, and the parameters are too complicated for the multi-block loading or spectrum loading.

4.3.2. Nonlinear Continuous Damage Rule

Based on the original concepts of Kachanov (1967) and Rabotnov (1969) in treating creep damage problems, continuum damage mechanics (CDM) based fatigue damage rules were proposed. A comprehensive review of cumulative fatigue damage theories for metals and alloys can be found in the literature (Fatemi and Yang 1998). The mechanical behaviour of a deteriorating medium at the continuum scale can be handled and fatigue damage in the region of fatigue crack initiation and growth of cracks in micro-scale can be well described. By measuring the changes in the tensile load-carrying capacity and using the effective stress concept, the nonlinear continuous fatigue damage model was expressed as the following equation with the damage rate dD being expressed in terms of cycles N (Chaboche and Lesne 1988):

$$dD = D^{\alpha(\sigma_M, \bar{\sigma})} \left[\frac{\sigma_M - \bar{\sigma}}{M_0(1 - b\bar{\sigma})} \right]^{\beta} dN \tag{4-16}$$

where M_0, b and β are material constants and α is a function of the stress state, $\bar{\sigma}$ is the mean stress, and σ_M is the maximum stress. This damage model is highly nonlinear in damage evolution and is able to include the mean stress effect. Based on the CDM concept, many forms of fatigue damage equations have been developed.

More recently, the CDM based fatigue damage models have been used in the fatigue analysis of long-span bridges due to the dynamic effects from strong wind, vehicle, train or their combined loads (Li et al. 2002, Xu et al. 2009). Despite the different proposed damage functions, the basic idea of fatigue damage accumulation rules is to calculate the fatigue damage in an evolutionary manor using a scalar damage variable (Liu and Mahadeven 2007).

In the present study, a nonlinear continuous fatigue damage model is used for fatigue damage assessment. The fatigue damage model was proposed by Chaboche and Lesne (1988) and named as nonlinear continuous damage rule (NLCDR), which is supported by Continuum Damage Mechanics and generalizes the model of Marco and Starkey (1954) and the Damage curve approach of Manson (1981).

It has been verified that the strain history of bridges under normal traffic can be approximately represented by a repeated block of cycles in which the cycles are daily repeated. Therefore, it is appropriate to use a stress cycle block to analyze and predict the bridge's fatigue life (Li et al. 2002). Predicting fatigue damage under a block-program can be expressed per the following recurrence formula (Chaboche and Lesne 1988). At the ith stress block, the damage accumulation parameter Y_i is defined as:

$$Y_i = D_i^{1-\alpha_i} \tag{4-17}$$

where D_i is the value of damage at the end of the ith block, with the associated value α_i for α. The function α is specified in the following way (Chaboche 1981):

$$\alpha_i = 1 - a \left\langle \frac{\sigma_{Mi} - \sigma_1(\bar{\sigma})}{\sigma_u - \sigma_{Mi}} \right\rangle \tag{4-18}$$

where $\sigma_1(\bar{\sigma})$ is the fatigue limit for a non-zero mean-stress, σ_u is the ultimate tensile strength, a is a coefficient depending on the material, and σ_{Mi} is the maximum stress in the block. The McCauley brackets symbol $<\,>$ is defined as $<u> = 0$ if $u < 0$ and $<u> = u$ if $u > 0$. When the maximum stress is lower than the fatigue limit $\sigma_1(\bar{\sigma})$, α equals to 1. In the block i, n_i cycles are applied. In the case of loadings above the fatigue limit, the fatigue damage accumulation parameter can be obtained:

$$Y_i = D_{i-1}^{1-\alpha_i} + \frac{n_i}{N_{F_i}} = \left(Y_{i-1} \right)^{\frac{1-\alpha_i}{1-\alpha_{i-1}}} + \frac{n_i}{N_{F_i}} \tag{4-19}$$

If the loadings are under the fatigue limit, the value of the damage at the end of the ith block is:

76

$$D_i = D_{i-1} \exp(n_i / N_i^*) \tag{4-20}$$

where N_i^* denotes a fictitious reference number of cycles to failure.

$$N_i^* = \left[\frac{\sigma_{Mi} - \bar{\sigma}_i}{M(\bar{\sigma}_i)} \right]^{-\beta} \tag{4-21}$$

where $M(\bar{\sigma}_i) = M_0(1 - b\bar{\sigma}_i)$ is the fatigue limit, $M_0 = \sigma_M N_f^{-3}$, $b=-0.1$ and $\beta=3$ are coefficients depending on the material (Chaboche and Lesne 1988).

When the fatigue damage variable D increases to 1, a fatigue failure is expected. In the present study, the failure function for fatigue (Limit State Function, LSF) is written as (Nyman and Moses 1985):

$$g = D_f - D(t) \tag{4-22}$$

where $D(t)$ is accumulated damage at time t and can be calculated via Eqs. (4-17) to (4-21); and g is a failure function such that $g<0$ implies a fatigue failure, D_f is the damage to cause failure and is treated as a random variable and assumed to follow a lognormal distribution with a mean value of 1 and a coefficient of variant (COV) of 0.15. The COV value is chosen to ensure that 95% of variable amplitude loading tests have a life within 70-130% (± 2 sigma) of the Miner's rule prediction (Nyman and Moses 1985).

4.3.3. Progressive Fatigue Reliability Assessment Approach

During the design life of bridges, the road roughness conditions deteriorated with each repeated block of stress cycles induced by multiple vehicle passages. The vehicle types, numbers, and distributions might change with time, as well. In the present study, a progressive fatigue reliability assessment approach is proposed and shown in Fig. 4-1. Multiple random variables in the vehicle-bridge dynamic system during the bridge's life cycle are included.

At each block of stress cycles, the vehicle types, numbers, and speeds are generated randomly according to their distributions. Five road roughness classifications were defined by the International Organization for Standardization (1995), and the ranges for the road roughness coefficients (RRC) were listed in Table 4-1. The road roughness coefficient for the current block of stress cycles is calculated based on the corresponding traffic information or can be adopted from the measured RRC records for existing bridges. In order to save calculation cost, the calculated or measured RRC is classified into one of the five classifications, and the vehicle-bridge dynamic analysis is carried out based only on the five classifications of RRC. If the PRC exceeds the maximum values for very poor conditions (2.048×10^{-3}), a surface renovation is expected. If that is the case, the road surface condition is re-assessed, and the road roughness condition will most likely be "very good" and deteriorate again as time goes.

Table 4-1 RRC values for road roughness classifications

Road roughness classifications	Ranges for RRCs
Very good	2×10^{-6}- 8×10^{-6}
Good	8×10^{-6}- 32×10^{-6}
Average	32×10^{-6} -128×10^{-6}
Poor	128×10^{-6} - 512×10^{-6}
Very poor	512×10^{-6} - 2048×10^{-6}

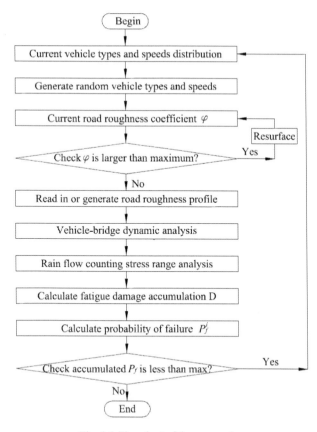

Fig. 4-1 Flowchart of the approach

Based on the road roughness coefficient, the road roughness profile is generated randomly using Eq. (4-9). After the vehicle-bridge dynamic analysis, the stress histories in bridge details are obtained, and rain-flow counting methods are used to calculate the numbers and magnitudes of the stress ranges. Based on the previous study, the revised equivalent stress ranges can be assumed to follow a normal or lognormal distribution (Zhang and Cai 2011). As a result, the revised stress range histories in the block of stress cycles can be randomly generated, and fatigue damage D_i is calculated using Eqs. (4-17) to (4-21). Based on the defined LSF in Eq. (4-22), a conditional probability of failure after the fatigue damage accumulation of the present block of stress cycles is obtained and recorded. The total accumulated probability of failure due to all of the preceding blocks of stress cycles can be calculated and compared with the maximum allowable value of probability of failure corresponding to the target reliability index, such as $\beta = 3.5$ in AASHTO (2010). If the accumulated probability of failure is less than the maximum allowable value, the analysis will continue to the next block of stress cycles. Otherwise, the cycle will stop, and the fatigue life for the target reliability index or fatigue reliability for a given design life of the bridges can be obtained.

4.4 Numerical Example

4.4.1. Prototype of Bridge and Vehicles

To demonstrate the proposed progressive fatigue damage prediction approach, a short span slab- on-girder bridge designed in accordance with AASHTO LRFD bridge design specifications (AASHTO 2010) is analyzed. The bridge, with a span length of 12 m and a width of 13m, accommodates two vehicle lanes traveling in the same direction. The concrete deck is 0.19m thick, and the haunch is 40mm high. All of the six steel girders are W27×94 and have an even spacing of 2.3m as shown in Fig. 4-2. Two intermediate and two end cross-frames enable the girders to deflect more equally. In this bridge, a steel channel section, C15×33.9, is used as a cross-frame. The fundamental frequency of the bridge is 14.5 Hz. The damping ratio is assumed to be 0.02. As a demonstration, the present study focuses on the fatigue analysis at the longitudinal welds located at the conjunction of the web and the bottom flange at the mid-span.

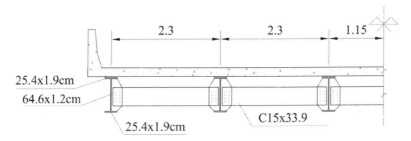

Fig. 4-2 Typical section of bridge (unit= meter)

In order to get the actual truckload spectra, weigh-in-motion (WIM) methodologies have been developed and are extensively used worldwide. Based on the data from WIM measurements, fifteen vehicle types are defined according to the FHWA classification scheme "F". Types five, eight and nine, representing the typical trucks with axle numbers of two, three and five, are predominant according to traffic data in the WIM stations in Florida (Wang and Liu 2000). In the present study, three-dimensional mathematic models of trucks are used, and the average daily truck traffic for the truck with two, three and five axles are assumed to be 600, 400, and 1000. Due to the small length of the bridge, only one truck is assumed passing the whole bridge at one time. The distributions of the vehicle speed are assumed to be the same for all the three types of vehicles.

The AASHTO H20-44, HS20-44 and 3S2 are used in the present study to represent the trucks with two, three and five axles as shown in Figs. 4-3 to 4-5, respectively. The geometry, mass distribution, damping, and stiffness of the tires and suspension systems of this truck are listed in Tables 4-2 to 4-4 (Zhang and Cai 2011). It is noteworthy that the design live load for the prototype bridge is HS20-44 truck. The purpose of using the three types of trucks in the present study is to make a comparison and investigate their effects on the fatigue life estimation. A 6m long approach slab connecting the pavement and bridge deck is considered.

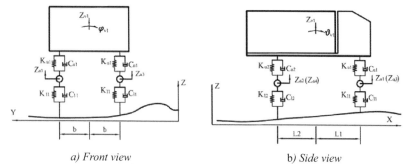

a) Front view   *b) Side view*

Fig. 4-3 Vehicle model for two axles

Table 4-2 Major parameters of a vehicle (2 axles)

	truck body	15233 kg
Mass	first axle suspension	725 kg
	second axle suspension	725 kg
Moment of inertia	Pitching, truck body1	19373 kg.m^2
	Rolling, truck body2	57690 kg.m^2
	Upper, 1st axle	242604 N/m
Spring stiffness	Lower , 1st axle	875082 N/m
	Upper, 2nd axle	1903172 N/m
	Lower , 2nd axle	3503307 N/m
	Upper, 1st axle	1314 N.s/m
Damping coefficient	Lower , 1st axle	2000 N.s/m
	Upper, 2nd axle	7445 N.s/m
	Lower , 2nd axle	2000 N.s/m
	L1	3.41 m
Length	L2	0.85 m
	B	1.1 m

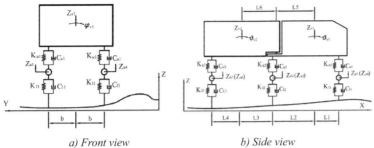

a) Front view *b) Side view*

Fig. 4-4   Vehicle model for three axles

Table 4-3 Major parameters of a vehicle (3 axles)

Mass	truck body 1	2612 kg
	truck body 2	26113 kg
	first axle suspension	490 kg
	second axle suspension	808 kg
	third axle suspension	653 kg
Moment of inertia	Pitching, truck body1	2022 kg.m^2
	Pitching, truck body2	33153 kg.m^2
	Rolling, truck body2	8544 kg.m^2
	Rolling, truck body2	181216 kg.m^2
Spring stiffness	Upper, 1st axle	242604 N/m
	Lower , 1st axle	875082 N/m
	Upper, 2nd axle	1903172 N/m
	Lower , 2nd axle	3503307 N/m
	Upper, 3rd axle	1969034 N/m
	Lower , 3rd axle	3507429 N/m
Damping coefficient	Upper, 1st axle	2190 N.s/m
	Lower , 1st axle	2000 N.s/m
	Upper, 2nd axle	7882 N.s/m
	Lower , 2nd axle	2000 N.s/m
	Upper, 3rd axle	7182 N.s/m
	Lower , 3rd axle	2000 N.s/m
Length	L1	1.698 m
	L2	2.569 m
	L3	1.984 m
	L4	2.283 m
	L5	2.215 m
	L6	2.338 m
	B	1.1 m

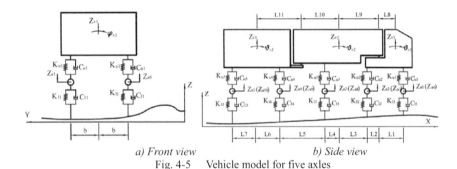

a) Front view   b) Side view

Fig. 4-5   Vehicle model for five axles

Table 4-4 Major parameters of a vehicle (5 axles)

	truck body 1	4956 kg
	truck body 2 & 3	20388 kg
Mass	first axle suspension	297 kg
	2nd & 3rd axle suspension	892 kg
	4th & 5th axle suspension	1054 kg
	Pitching, truck body1	3836 kg.m^2
Moment of inertia	Pitching, truck body2&3	20296 kg.m^2
	Rolling, truck body1	12291 kg.m^2
	Rolling, truck body2&3	333875 kg.m^2
	Upper, 1st axle	485208 N/m
	Lower , 1st axle	1402724 N/m
Spring stiffness	Upper, 2nd & 3rd axle	1396068 N/m
	Lower , 2nd & 3rd axle	5610546 N/m
	Upper, 4th & 5th axle	1359634 N/m
	Lower, 4th & 5th axle	5610546 N/m
	Upper, 1st axle	2400 N.s/m
	Lower , 1st axle	1600 N.s/m
Damping coefficient	Upper, 2nd & 3rd axle	7214 N.s/m
	Lower , 2nd & 3rd axle	1600 N.s/m
	Upper, 4th & 5th axle	7574 N.s/m
	Lower, 4th & 5th axle	1600 N.s/m
	L1	3 m
	L2	5 m
	L3	1.64 m
	L4	3.36 m
	L5	2.0 m
	L6	3.055 m
Length	L7	1.945 m
	L8	2.4 m
	L9	1.64 m
	L10	3.36 m
	L11	5.05 m
	B	1.1 m

After solving the equations of motions for the vehicle-bridge dynamic system, the revised equivalent stress ranges are obtained for varied vehicle speeds and road roughness conditions. The results for those cases with or without road surface discontinuities are obtained and shown in Figs. 4-6 to 4-8. Generally speaking, the revised equivalent stress ranges increase as the vehicle speed increases or the road surface condition deteriorates. When there are no surface discontinuities, S_w for 2-axle trucks remains almost the same for varying vehicle speed while it increases when the road surface condition deteriorates. When there are surface discontinuities, an increase trend of S_w for 2-axle trucks can be found as the vehicle speed increases. For 3-axle trucks, the increase trend of S_w is the same as the cases for 2-axle trucks. For 5-axle trucks, S_w remains almost the same for certain road surface conditions and no obvious increase trend can be found as the vehicle speed increases. However, S_w still keeps increasing with the deterioration of the road surface condition. The revised equivalent stress ranges greatly affect the fatigue damage assessment and fatigue life estimation that will be introduced in the later sections.

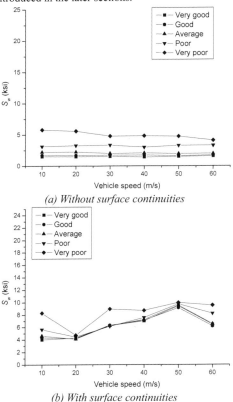

(a) Without surface continuities

(b) With surface continuities

Fig. 4-6   Equivalent stress ranges for 2 axle trucks

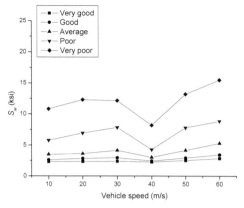

(a) Without surface continuities

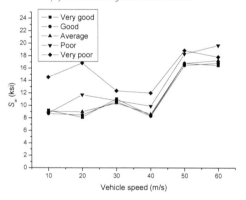

(b) With surface continuities
Fig. 4-7 Equivalent stress ranges for 3 axle trucks

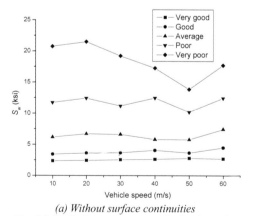

(a) Without surface continuities
Fig. 4-8 Equivalent stress ranges for 5 axle trucks

(Fig. 4-8 continued)

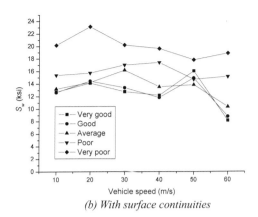

(b) With surface continuities

Nevertheless, the dynamic displacement of bridges was found to be changing with the vehicle speed in the literature (Green 1990, Paultre et al. 1992, Cai and Chen 2004; Cai et al. 2007). Typically, the maximum speed limits posted in bridges or roads are based on the 85th percentile speed when adequate speed samples are available. The 85th percentile speed is a value that is used by many states and cities for establishing regulatory speed zones (Donnell et al. 2009; TxDOT 2006). Statistical techniques show that a normal distribution occurs when random samples of traffic are collected. This allows describing the vehicle speed conveniently with two characteristics, i.e. the mean and standard deviation. In the present study, the 85th percentile speed is approximated as the sum of the mean value and one standard deviation for simplification. In the normal design condition, the speed limit is assumed as 26.8m/s (60mph), and the coefficient of variation of vehicle speeds is assumed as 0.2, which leads to a mean vehicle speed of 22.3 m/s (50mph).

4.4.2. Effect of Fatigue Damage Rules

Based on the proposed progressive fatigue reliability assessment approach, the fatigue life can be obtained based on LDR and NLCDR, respectively. Since the stress range histories in each block of stress cycles are generated randomly, which might lead to different fatigue life estimations, the mean values and standard deviations of fatigue life are obtained based on twenty stress histories. It is found that the coefficients of variance of the fatigue life from both models have the same value of 0.02.

For the purpose of demonstration, the curves for fatigue damage evolution and cumulative probability of failure from one randomly generated stress range history are shown in Fig. 4-9 and 4-10. When the LDR model is used, the fatigue damage index increases in a zigzag pattern due to the periodical road condition changes (road surface deterioration and renovation). The faster increase of fatigue damage corresponds to a deteriorated road condition (very bad) and the slower increase corresponds to a better road condition, for instance, a very good or good road condition after renovation. When the NLCDR model is used, the nonlinear effects on fatigue damage accumulation can be shown clearly in Fig. 4-9. Comparing the fatigue accumulation equation of Eq. (4-13) used by the LDR model and Eqs. (4-19) and (4-20) used by the NLCDR model, the fatigue damage accumulation is much smaller for the NLCDR model when the fatigue damage accumulation D_i remains small in early fatigue life period. Later, as the fatigue damage continues to accumulate and increases to a certain magnitude, the nonlinear effects of fatigue damage rule lead to a quicker fatigue damage accumulation when the NLCDR model is used. As a result, the fatigue damage index remains small in the first 110 years and increases fast thereafter in the present case study. The two curves for fatigue damage index intersect at the year of 133. In the first 133 years, the fatigue damage index obtained via the LDR model is larger than that obtained via the NLCDR model.

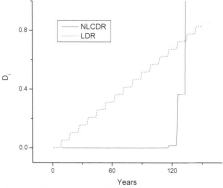

Fig. 4-9 Comparison of fatigue damage evolution

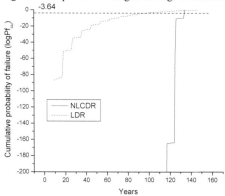

Fig. 4-10   Comparison of cumulative probability of failure

In the present study, the fatigue life is obtained based on the cumulative probability of failure. The target reliability index of 3.5, corresponding to the cumulative probability of failure of 2.3×10^{-4}, is used in the limit state function to define the fatigue failure. As shown in Fig. 4-10, the mean fatigue life of 133 years obtained from NLCDR is larger than the fatigue life of 96 years obtained from LDR. However, if the target reliability index is changed or the definition of failure is changed, for example, a different target reliability index, the fatigue life obtained from NLCDR model might be shorter than that obtained from the LDR model.

A bridge's fatigue life varies with multiple parameters, for instance, surface discontinuities, vehicle speed limit and its coefficient of variation. The effects of these parameters on fatigue lives are discussed in the following sections.

4.4.3.   Effects of Surface Discontinuities

As discussed earlier, a twofold road surface condition is used to include the randomly generated road profile from the zero-mean stationary Gaussian random process and the surface discontinuities, such as deck joints, cracks, potholes and delaminations. During each year in a bridge's life cycle, the surface discontinuities might exist for several days or months. In those days with surface discontinuities, the twofold road surface condition is used to generate the random road profiles, and the dynamic stress range differs from that of a road profile without surface discontinuities.

In order to study the effects of surface discontinuities on fatigue life estimations, six cases are defined as SD01 to SD06 based on the days of surface discontinuities in each year of the bridge's life cycle. From SD01 to SD06, the days with surface discontinuities are defined as 0, 1, 7, 15, 30 and 60 days. The calculated fatigue lives for various days with surface discontinuities are shown in Fig. 4-11. The surface discontinuities increase the stress ranges. As a result, the fatigue lives drop with the increase of days with discontinuities. In the current case study, the fatigue lives drop about 50% based on both the LDR and NLCDR models from cases SD01 to SD06.

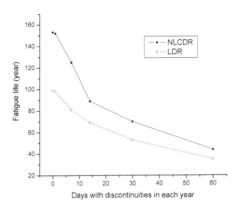

Fig. 4-11   Fatigue life estimation for SD cases

Comparison results of fatigue damage evolution and cumulative probability of failure for the six cases are shown in Figs. 4-12 and 4-13. Based on the NLCDR model, it takes 105 years for the damage index D_i to increase to 10^{-4} for Case SD-01 when there are no surface discontinuities as shown in Fig 4-12(a). With the increase of the days with surface discontinuities, the fatigue accumulations increase to the same amount much earlier. It takes only 27 years for the fatigue damage index D_i to increase to the same value in Case SD06 when there are 60 days of surface discontinuities in each year. For all the six cases from SD01 to SD06, the fatigue damage index D_i increase linearly in a zigzag pattern when the LDR model is used as shown in Fig. 4-12 (b). The slope of the curve increases with the number of days with surface discontinuities. In addition, the differences of the fatigue damage evolution curves for the six cases are smaller when the LDR model is used. For example, when the LDR model is used, it takes 18.1, 18.1, 17.5, 16.3, 8.9, and 8.2 years for the fatigue damage index D_i to increase to 0.1 for cases SD01 to SD06. The maximum year difference of the six cases is 10 years. However, when the NLCDR model is used, it takes 120.0, 142.5, 112.4, 77.6, 62.6, and 40.1 years for fatigue damage index D_i to increase to 0.1 for cases SD01 to SD06. The maximum year difference of the six cases is 80 years. Compared with the cases when the LDR model is used, the fatigue damage index of the NLCDR model is more sensitive to the number of days with surface discontinuities.

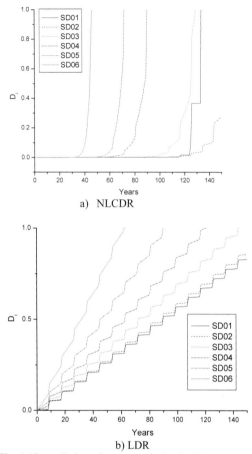

a) NLCDR

b) LDR

Fig. 4-12   Fatigue damage evolution for SD cases

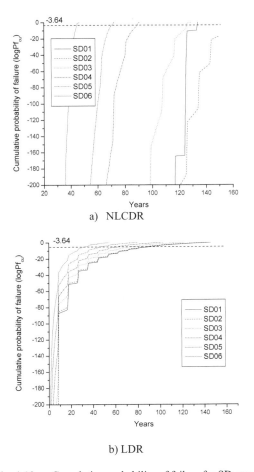

a) NLCDR

b) LDR

Fig. 4-13   Cumulative probability of failure for SD cases

The same trend can be found for the cumulative probability of failure as shown in Fig. 4-13. Based on the NLCDR model, it takes about 116 years for the cumulative probability of failure to increase to 10^{-200} for Case SD01 when there are no surface discontinuities. At the year of 134, the probability of failure exceeds 2.3×10^{-4} that corresponds to the target reliability index of $\beta=3.5$. The slopes of the curves for the cumulative probability of failure remain almost the same for the six cases as shown in Fig. 4-13 (a). However, when the LDR model is used, the slopes of the curves for the cumulative probability of failure decrease drastically after the first 20 years of fast increase of the cumulative probability of failure as shown in Fig. 4-13(b). The differences of the cumulative probability of failure to increase to the same magnitude, for instance, 10^{-60}, for different cases of SD01 to SD06 are less than 20 years in the first 40 years as shown in Fig. 4-13(b). As the cumulative probability of failure exceeds 2.3×10^{-4}, the time difference of fatigue lives increases to 89 years.

Based on the comparison results, the fatigue damage obtained from the LDR model is larger than that from the NLCDR model in the early life of bridges. However, the fatigue damage from the

NLCDR model develops much faster than that from the LDR model. In the present case, the fatigue lives obtained from the LDR model are less than the NLCDR model for different road conditions with varying days of road surface discontinuities. However, if the target reliability index is changed, the fatigue lives obtained from the LDR model might be larger than the NLCDR model due to their different curves of fatigue damage accumulation as shown in Figs. 4-9 and 4-10.

4.4.4. Effects of Vehicle Speed

Two parameters are used to define the vehicle speed distribution in the present study. One is the vehicle speed limit, and the other is its coefficient of variation. With the speed limit varying from 22.4m/s (50 mph) to 31.3m/s (70 mph) and the same coefficient of variation of 0.2, the obtained fatigue lives are shown in Fig. 4-14 (a) without surface discontinuities and Fig. 4-14 (b) with 15 days of surface discontinuities each year. When the speed limit is 22.4m/s (50mph), the vehicle speed range is around 20m/s. At such a vehicle speed as shown in Figs. 4-6 to 4-8, the variation of the revised equivalent stress ranges is large for the cases without surface continuities and small for the cases with surface continuities. As a result, at the speed limit of 22.4m/s (50mph), the fatigue life has a large variation for the cases without surface discontinuities and a small variation for cases with surface discontinuities. In addition, the revised equivalent stress ranges decrease with the increase of the vehicle speed, in several cases, when the vehicle speed increases from 20m/s to 30m/s as shown in Figs. 4-6 to 4-8. Consequently, the fatigue life increases with the vehicle speed limit as shown in Fig. 4-14(a). The dynamic responses of bridges are also affected by vehicle vibration frequencies (i.e., the vehicle suspension system) and vehicle speed induced resonant vibration effects, leading to a resonant vibration type of peak. In other words, the bridge vibration does not always monolithically increase with the increase of the vehicle speed (Shi et al. 2008).

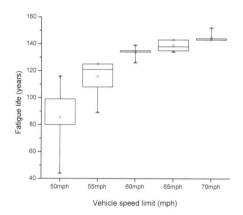

(a) Without surface discontinuities (NLCDR)

Fig. 4-14   Fatigue lives for varied speed limit

90

(Fig. 4-14 continued)

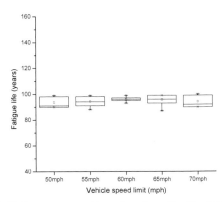

(b) 15 days per year with surface discontinuities (NLCDR)

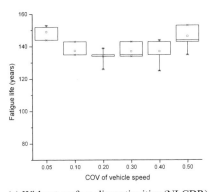

(a) Without surface discontinuities (NLCDR)

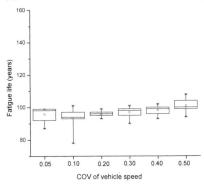

(b) 15 days per year with surface discontinuities (NLCDR)

Fig. 4-15   Fatigue lives for varied COV

91

With the same speed limit of 26.8m/s (60 mph) and the COV of speed being varied from 0.05, 0.10, 0.20, 0.30 to 0.04, the obtained fatigue lives are shown in Fig. 4-15 (a) without surface discontinuities and Fig. 4-15 (b) with 15 days of surface discontinuities each year. In Fig. 4-15(a), the fatigue lives are about the same years for different COVs. The fatigue lives reach their minimum when the COV is 0.2. When there are 15 days of surface discontinuities each year as shown in Fig. 4-15 (b), there are only very limited differences of fatigue lives for varying COV of the vehicle speed.

4.5 Concluding Remarks

During a bridge's life cycle, multiple dynamic loads can induce variant stress range cycles, and fatigue damage might accumulate and induce serious fatigue failure issues. During the most part of bridges' fatigue lives, the structure materials are in a linear range and micro cracks have not developed into macroscopic cracks. The fatigue life assessment of existing bridges is related to a sequence of progressive fatigue damage with only the initiations of micro cracks. Nonlinear cumulative fatigue damage theories were used to model the fatigue damage accumulation in this stage of the initiation of micro cracks. It is more appropriate to use the nonlinear continuous fatigue damage model for the fatigue analysis during a large fraction of bridges' life cycles.

This paper presents a progressive fatigue reliability assessment approach based on a nonlinear continuous fatigue damage model to include multiple random variables in the vehicle-bridge dynamic system during the bridge's life cycle. Multiple random variables for fatigue life estimations are included, for instance, vehicle speeds, vehicle types, and road surface profiles corresponding to progressively deteriorated road roughness conditions. The fatigue lives and fatigue damage index are obtained and compared with the results obtained from a linear fatigue damage model, as well. A fully computerized approach toward solving a coupled vehicle-bridge system including a 3-D suspension vehicle model and a 3-D dynamic bridge model is used to obtain the stress range histories. From the present study, the following conclusions are drawn:

1. The proposed approach is effective to predict the progressive fatigue reliability of existing bridges. Different fatigue damage models and various random variables of the vehicle-bridge dynamic system in a bridge's life cycle can be included in the proposed approach.

2. Significant discrepancies of fatigue damage estimations from the NLCDR model and LDR model are found. The fatigue damage estimated by using the LDR model is larger than that estimated by the NLCDR model in the early stage in bridges' life cycle. However, as the fatigue damage begins to accumulate, the fatigue damage increase rate of the NLCDR model is much faster than the LDR model.

3. Vehicle speeds have limited effects on the fatigue reliability and life while the days of road surface discontinuities have a large effect on the fatigue reliability and life.

Based on the proposed progressive fatigue reliability assessment approach, it is possible to predict bridge's progressive fatigue damage accumulation of existing bridges considering the effects from multiple random variables. Since the fatigue damage is calculated in a progressive manner, any unexpected damages due to overloads or some other dynamic impacts can be included in the stress range history to obtain the real fatigue life. In addition, the effects from the mean stress and maximum stress on fatigue life estimation are included in the nonlinear continuous damage model. In the present case, the fatigue life obtained from the NLCDR model is less than the results from LDR model. However, in order to ensure the structure's safety, it is necessary to calibrate the fatigue damage estimation and fatigue life of the bridge details due to dynamic loads using the frequently used the LDR model and the NLCDR model, taking advantage of the data from on-site experimental tests.

4.6 References

American Association of State Highway and Transportation Officials (AASHTO). (2010). LRFD bridge design specifications, Washington, DC.

Arnold, S. M., and Kruch, S. (1994). "A Differential CDM Model for Fatigue of Unidirectional Metal Matrix Composites." *International Journal of Damage Mechanics*, 3(2), 170-191.

Blejwas, T. E., Feng, C. C., and Ayre, R. S. (1979). "Dynamic interaction of moving vehicles and structures." *Journal of Sound and Vibration*, 67, 513-521.

Cai, C. S., and Chen, S. R. (2004). "Framework of vehicle-bridge-wind dynamic analysis." *Journal of Wind Engineering and Industrial Aerodynamics*, 92(7-8), 579-607.

Cai, C. S., Shi, X. M., Araujo, M., and Chen, S. R. (2007). "Effect of approach span condition on vehicle-induced dynamic response of slab-on-girder road bridges." *Engineering Structures*, 29(12), 3210-3226.

Cebon, D. (1999). Handbook of vehicle-road interaction, 2nd Ed., Swets & Zeitlinger B.V., Lisse, the Netherlands.

Chaboche, J. L., and Lesne, P. M. (1988). "A Non-Linear Continuous Fatigue Damage Model." *Fatigue & Fracture of Engineering Materials & Structures*, 11(1), 1-17.

Chen, Z. W., Xu, Y. L., Li, Q., and Wu, D. J. (2011). "Dynamic Stress Analysis of Long Suspension Bridges under Wind, Railway, and Highway Loadings." *J. Bridge Eng.*, 16(3), 383.

Deng, L., and Cai, C. S. (2010). "Development of dynamic impact factor for performance evaluation of existing multi-girder concrete bridges." *Engineering Structures*, 32(1), 21-31.

Dodds, C. J., and Robson, J. D. (1973). "The Description of Road Surface Roughness." *Journal of Sound and Vibration*, 31(2), 175-183.

Donnell, E. T., Hines, S. C., Mahoney, K. M., Porter, R. J., and McGee, H. (2009). "Speed Concepts: Informational Guide." U.S. Department of Transportation and Federal Highway Administration. Publication No. FHWA-SA-10-001.

Fatemi, A., and Yang, L. (1998). "Cumulative fatigue damage and life prediction theories: a survey of the state of the art for homogeneous materials." *International Journal of Fatigue*, 20(1), 9-34.

Green, M. F. (1990). "The dynamic response of short-span highway bridges to heavy vehicle loads," Ph.D. Dissertation, University of Cambridge, Cambridge, UK.

Guo, W. H., and Xu, Y. L. (2001). "Fully computerized approach to study cable-stayed bridge-vehicle interaction." *Journal of Sound and Vibration*, 248(4), 745-761.

Halford, G. R. (1997). "Cumulative fatigue damage modeling - crack nucleation and early growth." *International Journal of Fatigue*, 19, 253-60.

International Standard Organization. (1995). "Mechanical vibration - Road surface profiles - Reporting of measured data." Geneva.

Kachanov, L. M. (1967). The Theory of Creep, National Lending Library for Science and Technology, Boston Spa, England, Chaps. IX, X.

Kawai, M., and Hachinohe, A. (2002). "Two-stress level fatigue of unidirectional fiber-metal hybrid composite: GLARE 2." *International Journal of Fatigue*, 24(5), 567-580.

Keating, P. B., and Fisher, J. W. (1986). "Evaluation of Fatigue Tests and Design Criteria on Welded Details." NCHRP Report 286, Transportation Research Board, Washington, D.C.

Kwon, K., and Frangopol, D. M. (2010). "Bridge fatigue reliability assessment using probability density functions of equivalent stress range based on field monitoring data." *International Journal of Fatigue*, 32(8), 1221-1232.

Li, Z. X., Chan, T. H. T., and Ko, J. M. (2002). "Evaluation of typhoon induced fatigue damage for Tsing Ma Bridge." *Engineering Structures*, 24(8), 1035-1047.

Liu, Y., and Mahadevan, S. (2007). "Stochastic fatigue damage modeling under variable amplitude loading." *International Journal of Fatigue*, 29(6), 1149-1161.

Manson, S. S., and Halford, G. R. (1981). "Practical implementation of the double linear damage rule and damage curve approach for treating cumulative fatigue damage." *International Journal of Fracture*, 17(2), 169-192.

Marco, S. M., and Starkey, W. L. (1954). "A concept of fatigue damage." *Transactions of the A.S.M.E.*, 76, 627-32.

Miller, J.S. and Bellinger, W.Y. "Distress Identification Manual for the Long-Term Pavement Performance Program (Fourth Revised Edition)" Federal Highway Administration, U.S. Department of Transportation, Publication NO. FHWA-RD-03-031, Jun 2003.

Miner, M. A. (1945). "Cumulative damage in fatigue." *Journal of Applied Mechanics*, 67, A159-64.

Nyman, W. E., and Moses, F. (1985). "Calibration of Bridge Fatigue Design Model." *Journal of Structural Engineering*, 111(6), 1251-1266.

Paultre, P., Chaallal, O., and Proulx, J. (1992). "Bridge dynamics and dynamic amplification factors - a review of analytical and experimental findings." *Canadian Journal of Civil Engineering*, 19(2), 260-278.

Rabotnov, Y. N. (1969). Creep Problems in Structural Members, North Holland, Amsterdam.

Shimokawa, T., and Tanaka, S. (1980). "A statistical consideration of Miner's rule." International Journal of Fatigue, 2(4), 165-170.

Shi, X., Cai, C. S., and Chen, S. (2008). "Vehicle Induced Dynamic Behavior of Short-Span Slab Bridges Considering Effect of Approach Slab Condition." *Journal of Bridge Engineering*, 13(1), 83-92.

TxDOT. (2006). "Procedures for Establishing Speed Zones." Texas Department of Transportation.

Wang, T. L., and Huang, D. (1992). "Computer modeling analysis in bridge evaluation." Florida Department of Transportation, Tallahassee, FL.

Wang, T. L, and Liu, C.H. (2000). "Influence of Heavy Trucks on Highway Bridges," Rep. No. FL/DOT/RMC/6672-379, Florida Department of Transportation, Tallahassee, FL.

Xu, Y. L., Liu, T. T., and Zhang, W. S. (2009). "Buffeting-induced fatigue damage assessment of a long suspension bridge." *International Journal of Fatigue*, 31(3), 575-586.

Yao, J. T. P., Kozin, F., Wen, Y. K., Yang, J. N., Schueller, G. I., and Ditlevsen, O. (1986). "Stochastic fatigue fracture and damage analysis." Structural Safety(3), 231-67.

Zhang, W., C. S. Cai. (2011). "Fatigue Reliability Assessment for Existing Bridges Considering Vehicle and Road Surface Conditions", Journal of Bridge Engineering, doi:10.1061/ (ASCE) BE. 1943-5592.0000272

CHAPTER 5 FINITE ELEMENT MODELING OF BRIDGES WITH EQUIVALENT ORTHOTROPIC MATERIAL METHOD FOR MULTI-SCALE DYNAMIC LOADS

5.1 Introduction

With an increase of span lengths, bridge structures are becoming more flexible, which makes them more vulnerable to the wind-induced vibrations or flutter failure at a low critical wind speed. The wind induced buffeting vibrations can produce repeated dynamic stresses in bridge details (Gu et al. 1999, Xu et al. 2009). Nevertheless, local vehicle dynamic loads can cause repeated dynamic stresses and induce local fatigue damages or cracks, as well. Such a local failure might develop and induce the whole structure failure, for instance, the collapse and failure of the King's Bridge in Melbourne, Australia (1962), the Point Pleasant Bridge in West Virginia (1967) and Yellow Mill Pond Bridge in Connecticut (1976).

For the purpose of fatigue damage predictions of long-span bridge details due to combined dynamic loads from vehicles and winds, several schemes are used to model the long-span bridges. Traditionally, a global structural analysis using a beam element model is first conducted to determine the critical locations and a local analysis is carried out to obtain the local effects. The global beam element model of long-span bridges is usually in kilometers, and the majority of finite element models in the previous studies are built using beam elements. Such a model is usually called as a "fish-bone" model (Chan et al. 2008). In the beam element model, the whole section is assumed to deform with respect to the centroid of the bridge deck system and all the mass and stiffness properties are transformed to the equivalent beams located along the centroid of each deck section. The equivalent beam forms the spine of the "fish-bone". Rigid beams are used to locate one end of the cables or hangers on the bridge decks for cable supported bridges, which form the ribs of the "fish-bone". The overall static and dynamic response can be obtained at each node located at each beam end. However, only the rigid body motion is considered in the plane of the bridge deck section and the local deformations are neglected.

Based on the St. Venant's principle, the localized effects from loads will dissipate or smooth out with regions that are sufficiently away from the location of the load (Mises 1945). The forces are obtained from the beam element model and implemented only on a portion of the overall geometry to obtain the local static effects (Wu et al. 2003). Large computation efforts are needed for the refined section model with complicated structural details, and it is difficult to include the time-varying dynamic effects from both wind loads and vehicle loads. Chan et al. (2005) merged a typical detailed joint geometry model into the beam element model to obtain the hot-spot stress concentration factors (SCF) of typical welded joints of the bridge deck. Then the hot spot stress block cycles were calculated by multiplying the nominal stress block cycles by the SCF for fatigue assessment. Li et al. (2007) proposed a multi-scale FE modeling strategy for long-span bridges. The global structural analysis was carried out using the beam element modeling method at the level of a meter. The local detailed hot-spot stress analysis was carried out using shell or solid elements at the level of a millimeter. After introducing the mixed dimensional coupling constraint equations developed by Monaghan (2000), the multi-scale model of the Tsing Ma Bridge was built and the computed results were obtained and verified using the on-site measurement data. However, due to the limitations of the beam element modeling, the effects from distortion, constrained torsion, and shear lag were missing in the previous analyses, which might have a large effect on the local displacements, strains, and stresses for wide bridge decks with weak lateral connections.

With the increasing demand of traffic, the bridge decks are becoming wider and have a large mass distribution across the bridge deck or even have a separate deck section type, such as the Stonecutters cable-stayed bridge and the Xihoumen suspension bridge in China. Therefore, it might not be reasonable to assume rigid body motion over the full bridge deck due to its weak lateral connections. Two or more parallel "fish-spines" are suggested for the beam element model to model the bridge deck with multiple centroids of separate decks in order to obtain a reasonable result (Du 2006). Nevertheless, in order to enhance the bending resistance of the steel plate to carry local loads from vehicle wheels, steel plates of the bridge decks are often stiffened with multiple closed or open stiffeners. As a result, it is impossible to numerically model the long span bridges with complicated structural details with a simple beam element model. The stress histories in structural details due to the dynamic effects from vehicle loads and wind loads cannot be obtained, either. Therefore, a multiple scale modeling scheme is essential to effectively model the structure in detail and save the calculation cost with less numbers of elements and nodes. Based on the principle of equivalent stiffness properties in both the lateral and longitudinal directions of the steel plate with multiple stiffeners, equivalent orthotropic shell elements were proposed to model the long-span bridges and the local deformation effects can be taken into account (Zhang and Ge 2003).

In the present study, a multiple scale finite element modeling scheme is presented based on the equivalent orthotropic material modeling (EOMM) method. Bridge details with multiple stiffeners are modeled with shell elements using an equivalent orthotropic material. The static and dynamic responses and dynamic properties of a simplified short span bridge from the EOMM shell element model are obtained and compared with the results from the original shell element model with its real geometry and materials. The EOMM shell element model for a long-span bridge is also built with good precisions on dynamic properties. The paper is organized as the following three main sections. In the first section, the equivalent orthotropic modeling method is introduced, followed by the control equations for the finite element analysis on static and dynamic performance and dynamic properties. In the second section, the vehicle-bridge dynamic system and its parameters are introduced. In the third section, two numerical examples are presented, including one short span bridge and one long span bridge. For the short span bridge, comparisons are made on static and dynamic analysis and dynamic properties between the EOMM shell element model and the original shell element model. For the long-span bridge, the dynamic properties are compared between the results from the EOMM shell element model and the beam element model. Conclusions are drawn from the case study results at the end of the paper.

5.2 Equivalent Orthotropic Material Modeling Method

5.2.1. Orthotropic Bridge Deck

Most of the metallic alloys and thermoset polymers are considered isotropic i.e., their properties are independent of directions. In their stiffness and compliance matrices, only two elastic constants, namely, the Young's modulus E and the Poisson's ratio v are independent. In contrast, the orthogonal materials have independent material properties in at least two orthogonal planes of symmetry. A total number of 21 elastic constants are needed for fully anisotropic materials without any plane of symmetry.

In order to enhance the bending resistance of the steel plate to carry local loads from vehicle wheels, orthotropic bridge decks were developed by German engineers in the 1950s (Wolchuk 1963). As a result, the total cross-sectional area of steel in the plate was increased, and the overall bending capacity of the deck and the resistance of the plate to buckling were increased, as well. The creative orthotropic bridge design not only offered excellent structural characteristics, but were also economical to build (Troisky 1987). From short span bridges to large span cable-supported bridges, the orthotropic bridge design was used throughout the world, for instance, the Golden Gate Bridge and the Akashi-Kaikyo Bridge. In addition to the bridge deck, the orthotropic steel plates are used in the other parts of the bridge deck systems, such as the cross plates or the side plates. For example, the Donghai Cable-stayed Bridge in China with a main span of 420m has a prestressed concrete deck, while the web, cross plates and bottom plates have multiple various open and closed rib

97

stiffeners. The bridge deck system is shown in Fig. 5-1 (Wu et al 2003). In order to model the bridges with small stiffeners, large computational efforts are needed if all the stiffeners are modeled in details, and it is almost impossible to carry out the corresponding dynamic analysis. Due to the orthotropic properties of the deck plate, it is possible to use equivalent orthotropic shell elements to model the plate with various stiffeners.

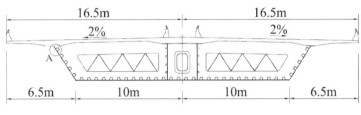

a)   Cross section

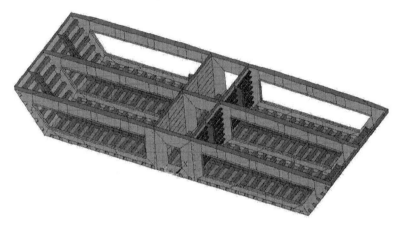

b)   Various open or closed ribs in bridge deck

Fig. 5-1 Bridge deck system of Donghai Bridge

5.2.2.   Orthotropic Shell Element

A typical shell element is subjected to both membrane forces and bending forces. The quadrilateral flat shell element can be assembled by the four node quadrilateral plane stress element and the quadrilateral plate bending element based on the discrete Kirchoff theory (DKQ) (Batoz and Thar 1982). For an x-y plane shell element, the assembly of the quadrilateral flat shell element can be represented as shown in Fig. 5-2 (Kansara 2004). The translations and rotations are represented by single and double arrows, respectively.

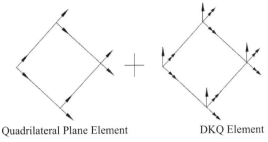

Quadrilateral Plane Element   DKQ Element

Fig. 5-2 Degrees of freedom of quadrilateral shell element

Each node of the shell element has six degrees of freedom and the corresponding nodal displacements are

$$\{U_i\} = \{u_i \quad v_i \quad w_i \quad \theta_{xi} \quad \theta_{yi} \quad \theta_{zi}\} \tag{5-1}$$

where u, v, w are the translations and θ_{xi}, θ_{yi}, θ_{zi} are the rotations in the x, y, and z direction, respectively.

For orthotropic materials, the Hooke's law in stiffness form can be given by:

$$\begin{bmatrix} \sigma_{xx} \\ \sigma_{yy} \\ \sigma_{zz} \\ \sigma_{yz} \\ \sigma_{zx} \\ \sigma_{xy} \end{bmatrix} = \begin{bmatrix} \dfrac{1-v_{yz}v_{zy}}{E_yE_z\Delta} & \dfrac{v_{yx}+v_{zx}v_{yz}}{E_yE_z\Delta} & \dfrac{v_{zx}+v_{yx}v_{zy}}{E_yE_z\Delta} & 0 & 0 & 0 \\[2mm] \dfrac{v_{xy}+v_{xz}v_{zy}}{E_zE_x\Delta} & \dfrac{1-v_{zx}v_{xz}}{E_zE_x\Delta} & \dfrac{v_{zy}+v_{zx}v_{xy}}{E_zE_x\Delta} & 0 & 0 & 0 \\[2mm] \dfrac{v_{xz}+v_{xy}v_{zy}}{E_xE_y\Delta} & \dfrac{v_{yz}+v_{xz}v_{yx}}{E_xE_y\Delta} & \dfrac{1-v_{xy}v_{yx}}{E_xE_y\Delta} & 0 & 0 & 0 \\[2mm] 0 & 0 & 0 & 2G_{yz} & & \\[1mm] 0 & 0 & 0 & & 2G_{zx} & \\[1mm] 0 & 0 & 0 & & & 2G_{xy} \end{bmatrix} \begin{bmatrix} \varepsilon_{xx} \\ \varepsilon_{yy} \\ \varepsilon_{zz} \\ \varepsilon_{yz} \\ \varepsilon_{zx} \\ \varepsilon_{xy} \end{bmatrix} \tag{5-2}$$

where $\Delta = \dfrac{1-v_{xy}v_{yx}-v_{yz}v_{zy}-v_{zx}v_{xz}-2v_{xy}v_{yz}v_{zx}}{E_xE_yE_z}$, E_x, E_y, E_z are the Young's modulus in x, y and z directions, respectively, and v_{ij} are the Poisson's ratios. The element stiffness matrix for the shell element is assembled by superimposing the membrane stiffness and bending stiffness matrices together:

$$\mathbf{k} = \mathbf{k_b} + \mathbf{k_m} \tag{5-3}$$

The element stiffness matrix for the quadrilateral plane element and the DKQ element can be obtained as (Kansara 2004)

$$[k]_{8\times8} = t\sum_{j=1}^{2}\sum_{i=1}^{2}w_jw_i\left[B(\xi_i,\eta_j)\right]_{8\times3}^T[E]_{3\times3}\left[B(\xi_i,\eta_j)\right]_{3\times8}\left|J(\xi_i,\eta_j)\right|d\xi d\eta \tag{5-4}$$

$$[k]_{12\times12} = \sum_{j=1}^{2}\sum_{i=1}^{2}w_jw_i\left[B(\xi_i,\eta_j)\right]_{12\times3}^T[D_b]_{3\times3}\left[B(\xi_i,\eta_j)\right]_{3\times12}\left|J(\xi_i,\eta_j)\right|d\xi d\eta \tag{5-5}$$

99

where $[k]_{8\times8}$ and $[k]_{12\times12}$ are the stiffness matrix for the two elements, E is the material matrix for in plane deformation, D is the material matrix for bending, B is the strain-displacement matrix for the shell element, and J is the Jacobian matrix that is used to convert the strain-displacement matrix from the element local coordinate system to the natural coordinate system. Therefore, the global stiffness matrix can be formed using the discretized element stiffness matrices.

5.2.3. Equivalent Orthotropic Material Modeling Method

In the longitudinal and lateral directions, the multiple open or closed ribs provide varied stiffness to the steel plates in the bridge deck, such as the longitudinal trapezoidal stiffeners shown in Fig. 5-1. In order to avoid unmanageably large numbers of elements and degrees of freedom involved in solving the equations of the motions of the bridge, an equivalent orthotropic shell element is used to model the steel plates with stiffeners. In the present study, the following equivalent rules for the equivalent orthotropic shell element are applied.

Firstly, the equivalent orthotropic shell element has the same bending stiffness in the unit width as the original configuration:

$$E_x \frac{d^3}{12} = EI_x, E_y \frac{d^3}{12} = EI_y \tag{5-6}$$

where d is the equivalent thickness of the shell, E_x and E_y are the equivalent elastic modulus in two orthogonal x and y directions, I_x and I_y are the moment of inertia in x and y directions, and E is the elastic modulus of the original material. The x direction is along the bridge, and y direction is perpendicular to x direction in the shell plane.

Secondly, the equivalent orthotropic shell has the same longitudinal stiffness and shear stiffness as the original configuration:

$$E_x d = EA, G_{xy} d = Gt \tag{5-7}$$

where A is the area of the unit width of the shell, G_{xy} is the equivalent shear modulus, t is the thickness of the plate, and G is the shear modulus of the original material.

Thirdly, the equivalent orthotropic shell element has the same weight as the original element:

$$\rho_e d = \rho A \tag{5-8}$$

where ρ_e is the equivalent density of the shell element and ρ is the original density of the structure material.

Based on the equivalent rules, the material properties of the equivalent shell element can be obtained from Eqs. (6) to (8):

$$d = \sqrt{\frac{12I_x}{A}}, E_x = EA\sqrt{\frac{A}{12I_x}}, E_y = \frac{12EI_y}{d^3} A, G_{xy} = \frac{Gt}{d}, \rho_e = \frac{\rho A}{d} \tag{5-9}$$

In addition, the locations of each equivalent shell element remain unchanged and the following assumptions are made in order to define the material matrix listed in Eq. (5-2). The Poisson's ratios v_{ij} are all assumed to be zeros and E_z, G_{yz} and G_{xz} have a relatively small value compared with other modulus in other directions.

Based on the equivalent orthotropic material modeling (EOMM) scheme, bridge components with complicated structural details such as multiple stiffeners are modeled as the equivalent shell element using the equivalent orthotropic material. The material matrix for the equivalent orthotropic material is obtained and the element stiffness matrix is changed accordingly as shown in Eq. (5-10).

$$[E]^{eq} = \begin{bmatrix} E\sqrt{\dfrac{A^3}{12I_x}} & 0 & 0 & 0 & 0 & 0 \\ 0 & \dfrac{12EI_y}{d^3} & 0 & 0 & 0 & 0 \\ 0 & 0 & E_z & 0 & 0 & 0 \\ 0 & 0 & 0 & 2G_{yz} & 0 & 0 \\ 0 & 0 & 0 & 0 & 2G_{zx} & 0 \\ 0 & 0 & 0 & 0 & 0 & 2\dfrac{Gt}{d} \end{bmatrix} \tag{5-10}$$

After using the EOMM method, the coupled effects between different directions disappear. Since the equivalence is only based on its longitudinal stiffness EI_x, lateral stiffness EI_y and shear stiffness $G_{xy}t$, some differences are expected for the output results related to the other elastic and shear moduli of E_z, G_{yz} and G_{zx}. Compared with the simple finite element model built with beam elements, the EOMM shell element model has a better modeling of the structural details and a better simulation of stiffness and mass distribution in the bridge deck sections.

5.2.4. Control Equation for FE Analysis

The control equations for dynamic structure systems are the equations of motions as shown in the following:

$$\mathbf{M\ddot{X} + C\dot{X} + KX = F} \tag{5-11}$$

where $\mathbf{M}$, $\mathbf{C}$ and $\mathbf{K}$ are the mass, damping, and stiffness matrices, respectively, and $\mathbf{F}$ is the force vector. After obtaining the matrices and the vector, control equations for the dynamic system can be solved to obtain the dynamic response $\mathbf{X}$.

However, if both matrix $\mathbf{M}$ and matrix $\mathbf{C}$ are zero, Eq. (5-11) degrades to the control equation for a static problem, and the displacement vectors can be obtained via the following equation and the strains and stresses can also be obtained.

$$\mathbf{KX = F} \tag{5-12}$$

If both matrix $\mathbf{C}$ and vector $\mathbf{F}$ are zero, the control equations change to an eigenvalue problem as shown below, and the equation can be used for modal analysis to obtain the dynamic properties of the structure, for instance, natural frequencies and mode shapes.

$$\mathbf{M\ddot{X} + KX = 0} \tag{5-13}$$

The static and dynamic responses and the dynamic properties can be obtained when Eqs. (11), (12) or (13) are solved. Due to the complexity of the stiffness matrices, it is difficult to present a simple correlation equation for the static and dynamic responses of the bridges between the EOMM shell element model and the shell element model with its real material properties. Instead, numerical examples are presented in the following sections to compare the static and the dynamic properties of bridges. The dynamic responses of the vehicle-bridge dynamic system from the EOMM model and the model with real geometry and material properties can also be obtained and compared with each other.

5.3 Modeling of Vehicle-Bridge Dynamic System

5.3.1. Equations of Motions for Vehicle-Bridge Dynamic System

The interactions between the bridge and vehicles are modeled as coupling forces between the tires and the road surface. The coupling forces were proven to be significantly affected by the vehicle speed and road roughness conditions and resulted in significant effects on the dynamic responses of short span bridges (Deng and Cai 2010, Shi et al. 2008, Zhang and Cai 2011). In the present study, the vehicle is modeled as a combination of several rigid bodies connected by several axle mass blocks, springs, and damping devices (Cai and Chen 2004). The tires and suspension systems are idealized as linear elastic spring elements and dashpots. The equation of motion for the vehicle and the bridge are listed in the following matrix form:

$$[M_v]\{\ddot{d}_v\}+[C_v]\{\dot{d}_v\}+[K_v]\{d_v\}=\{F_v^G\}+\{F_c\}$$ (5-14)

$$[M_b]\{\ddot{d}_b\}+[C_b]\{\dot{d}_b\}+[K_b]\{d_b\}=\{F_b\}$$ (5-15)

where the mass matrix $[M_v]$, damping matrix $[C_v]$, and stiffness matrix $[K_v]$ are obtained by considering the equilibrium of the forces and moments of the system; $\{F_v^G\}$ is the self-weight of the vehicle; $\{F_c\}$ is the vector of wheel-road contact forces acting on the vehicle; $[M_b]$ is the mass matrix, $[C_b]$ is the damping matrix; $[K_b]$ is the stiffness matrix of the bridge; and $\{F_b\}$ is wheel-bridge contact forces on the bridge and can be stated as a function of deformation of the vehicle's lower spring:

$$\{F_b\}=-\{F_c\}=[K_l]\{\Delta_l\}+[C_l]\{\dot{\Delta}_l\}$$ (5-16)

where $[K_l]$ and $[C_l]$ are the coefficients of the vehicle's lower spring and damper; and Δ_l is the deformation of the lower springs of the vehicle. The relationship among the vehicle-axle-suspension displacement Z_a, displacement of bridge at wheel-road contact points Z_b, deformation of lower springs of vehicle Δ_l, and road surface profile $r(x)$ is derived as:

$$Z_a=Z_b+r(x)+\Delta_l$$ (5-17)

$$\dot{Z}_a=\dot{Z}_b+\dot{r}(x)+\dot{\Delta}_l$$ (5-18)

where $\dot{r}(x)=(dr(x)/dx)\cdot(dx/dt)=(dr(x)/dx)\cdot V(t)$ and $V(t)$ is the vehicle velocity.

Therefore, the contact force F_b and F_c between the vehicle and the bridge is derived as:

$$\{F_b\}=-\{F_c\}=[K_l]\{Z_a-Z_b-r(x)\}+[C_l]\{\dot{Z}_a-\dot{Z}_b-\dot{r}(x)\}$$ (5-19)

5.3.2. Mode Superposition Techniques

After transforming the contact forces to the equivalent nodal force and substituting them into Eqs. (14) and (15), the final equations of motion for the coupled system are derived as the following (Shi et al. 2008):

$$\begin{bmatrix} M_b & \\ & M_v \end{bmatrix}\begin{Bmatrix} \ddot{d}_b \\ \ddot{d}_v \end{Bmatrix}+\begin{bmatrix} C_b+C_{bb} & C_{bv} \\ C_{vb} & C_v \end{bmatrix}\begin{Bmatrix} \dot{d}_b \\ \dot{d}_v \end{Bmatrix}+\begin{bmatrix} K_b+K_{bb} & K_{bv} \\ K_{vb} & K_v \end{bmatrix}\begin{Bmatrix} d_b \\ d_v \end{Bmatrix}=\begin{Bmatrix} F_{br} \\ F_{vr}+F_v^G \end{Bmatrix}$$ (5-20)

The additional terms C_{bb}, C_{bv}, C_{vb}, K_{bb}, K_{bv}, K_{vb}, F_{br} and F_{vr} in Eq. (5-20) are due to the expansion of the contact force in comparison with Eqs. (14) and (15). When the vehicle is moving along the bridge, the bridge-vehicle contact points change with the vehicle longitudinal position and the road roughness at the contact point.

As a large number of degrees of freedom (DOF) are involved, the bridge mode superposition technique is used to simplify the modeling procedure based on the obtained bridge mode shape $\{\Phi_i\}$ and the corresponding natural circular frequencies ω_i. Bridge fatigue analysis corresponds to service load level, and the bridge performance is practically in the linear range, which justifies the use of the modal superposition approach. Consequently, the number of equations in Eq. (5-20) and the complexity of the whole procedure are greatly reduced. The bridge dynamic response $\{d_b\}$ can be expressed as:

$$\{d_b\} = [\Phi_b]\{\xi_b\} = \big[\{\Phi_1\} \quad \{\Phi_2\} ... \{\Phi_n\}\big]\{\xi_1 \quad \xi_2 \cdots \xi_n\}^T \qquad (5\text{-}21)$$

where n is the total number of modes for the bridge under consideration, and $\{\Phi_i\}$ and ξ_i are the ith mode shape and its generalized coordinates. If each mode shape is normalized to the mass matrix, i.e. $\{\Phi_i\}^T[M_b]\{\Phi_i\}=1$ and $\{\Phi_i\}^T[K_b]\{\Phi_i\}=\omega_i^2$, and if the damping matrix $[C_b]$ is assumed to be $2\omega_i\eta_i[M_b]$, where ω_i is the natural circular frequency of the bridge and η_i is the percentage of the critical damping for the bridge i^{th} mode, Eq. (5-20) can be rewritten as (Shi et al. 2008):

$$\begin{bmatrix} I & \\ & M_v \end{bmatrix}\begin{Bmatrix} \ddot{\xi}_b \\ \ddot{d}_v \end{Bmatrix} + \begin{bmatrix} 2\omega_i\eta_iI + \Phi_b^TC_{bb}\Phi_b & \Phi_b^TC_{bv} \\ C_{vb}\Phi_b & C_v \end{bmatrix}\begin{Bmatrix} \dot{\xi}_b \\ \dot{d}_v \end{Bmatrix} + \begin{bmatrix} \omega_i^2I + \Phi_b^TK_{bb}\Phi_b & \Phi_b^TK_{bv} \\ K_{vb}\Phi_b & K_v \end{bmatrix}\begin{Bmatrix} \xi_b \\ d_v \end{Bmatrix} = \begin{Bmatrix} \Phi_b^TF_{br} \\ F_{vr}+F_v^G \end{Bmatrix} \qquad (5\text{-}22)$$

The stress vector can be obtained by:

$$[S] = [E][B]\{d_b\} \qquad (5\text{-}23)$$

where $[E]$ is the stress-strain matrix, and $[B]$ is the strain-displacement matrix assembled with x, y and z derivatives of the element shape functions.

Since multiple stress cycles might be found in the stress time histories, two correlated parameters are essential to evaluate the fatigue damages induced by the stress cycles, i.e. the equivalent stress range and the number of stress cycles due to each truck passage. For the vehicle-bridge dynamic system, cycle counting methods, such as the rainflow counting method, are used to obtain the number of cycles per truck passage. In order to make comparisons between the results from the original shell element model built with the real materials and the EOMM shell element model built with the equivalent orthotropic materials, a revised equivalent stress range S_w is used in the present study. The parameter, S_w, combines the two essential parameters, namely, the equivalent stress range and the number of cycles per truck passage, for fatigue damage evaluation (Zhang and Cai 2011). The fatigue damage of multiple stress cycles due to each truck passage is considered as the same as that of a single stress cycle of S_w. For each truck passage, the revised equivalent stress range is:

$$S_w = \left(N_c\right)^{1/m} \cdot S_{re} \qquad (5\text{-}24)$$

where N_c is the number of stress cycles due to the truck passage, S_{re} is the equivalent stress range of the stress cycles by the truck, and m is the material constant that could be assumed as 3.0 for all fatigue categories (Keating and Fisher 1986).

5.3.3. Simulation of Road Surface Roughness

Based on the studies carried out by Dodds and Robson (1973) and Honda et al. (1982), the long undulations in the roadway pavement could be assumed as a zero-mean stationary Gaussian random process, and it could be generated through an inverse Fourier transformation (Wang and Huang 1992):

$$r(x) = \sum_{k=1}^{N} \sqrt{2\phi(n_k)\Delta n} \cos(2\pi n_k x + \theta_k) \tag{5-25}$$

where θ_k is the random phase angle uniformly distributed from 0 to 2π; $\phi()$ is the power spectral density (PSD) function (m^3/cycle) for the road surface elevation; and n_k is the wave number (cycle/m). The PSD functions for road surface roughness were developed by Dodds and Robson (1973), and three groups of road classes were defined with the values of roughness exponents ranging from 1.36 to 2.28 for motorways, principal roads, and minor roads. In order to simplify the description of road surface roughness, both of the two roughness exponents were assumed to have a value of two and the PSD function was simplified by Wang and Huang (1992) as:

$$\phi(n) = \phi(n_0)(\frac{n}{n_0})^{-2} \tag{5-26}$$

where $\phi(n)$ is the PSD function (m^3/cycle) for the road surface elevation; n is the spatial frequency (cycle/m); n_0 is the discontinuity frequency of $1/2\pi$ (cycle/m); and $\phi(n_0)$ is the road roughness coefficient (m^3/cycle) whose value is chosen depending on the road condition. The International Organization for Standardization (1995) has proposed the road roughness classification from very good, good, average, and poor to very poor according to different values of $\phi(n_0)$.

5.4 Numerical Example

5.4.1. Prototype Bridge and Vehicle Model

In order to demonstrate and validate the EOMM method in predicting the static and dynamic responses of bridges, two bridges, namely, a simplified short span beam bridge and a long span cable-stayed bridge are analyzed.

The short span bridge has a square cross section with two closed form stiffeners installed on the bottom plate as shown in Fig. 5-3. The bridge has a span length of 10 m and a width of 2.4m, which can accommodate one vehicle traveling along the centerlines of the bridge deck. The thickness of the four side plates of the deck is 0.016m. The bottom plate has two closed rectangle stiffeners with a side width of 0.3m and a height of 0.26m. The thickness of the stiffeners is 0.008m. The distances from the two stiffeners to the side plates are 0.7m, and they are 0.4m apart from each other. The cross plates have a thickness of 0.016m, and they are 0.5m apart with each other in the longitudinal direction. The elasticity modulus of the steel used in the bridge deck is 2.1×10^{11} Pa, and the Poisson's ratio is 0.3.

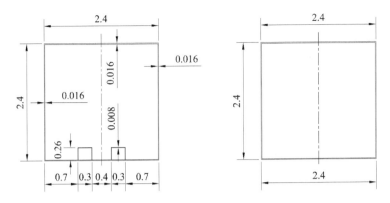

a) Real bridge deck section   b) Bridge deck with equivalent plate

Fig. 5-3 Bridge deck section of the short-span bridge

Based on the EOMM method, the bottom plate with stiffeners is equivalent to a flat plate with equivalent material properties. The other plates remain the same as the real ones. Therefore, the bridge model with the equivalent plate has the same geometry and the same nodes in the four corners of the section as the actual section. The equivalent material properties are obtained based on the EOMM method and listed in Table 5-1. The two meshed bridge models are meshed in the same size and are shown in Fig. 5-4. For a demonstration of the bridge's dynamic performance under vehicle loads, the HS20-44 truck is used as the prototype of the vehicle as shown in Fig. 5-5. The geometry, mass distribution, damping, and stiffness of the tires and suspension systems of the truck are listed in Table 5-2. The long-span Donghai Cable-stayed Bridge is also used for the demonstration in the present study. It has a main span of 420m and is located in a typhoon zone of east China. The deck of the girder is made of prestressed concrete, while the web, cross plates, and bottom plates with multiple stiffeners are made of steel as shown in Fig. 5-1.

Table 5-1 Material property for bottom plate of the short-span bridge

Category	Original material (steel)	Equivalent material using EOMM
Elasticity modulus	$E=2.1\times10^{11}$ Pa	$E_x = 1.33\times10^8$ Pa
		$E_y = 2.75\times10^7$ Pa
		$E_z = 2.10\times10^8$ Pa
Shear modulus	$G_{xy}=8.1\times10^{10}$ Pa	$G_{xy}=3.84\times10^8$ Pa
		$G_{yz}=3.84\times10^6$ Pa
		$G_{xz}=3.84\times10^6$ Pa
Poisson's ratio	$v = 0.3$	$v_{xx} = v_{xz} = v_{zy}=0$
Thickness of the plate	$t=0.016$m	$d = 3.368$m
Density	$\rho=7850$kg/m^3	$\rho^e= 49.877$kg/m^3

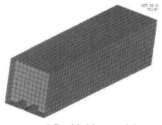 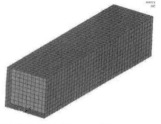

a) Real bridge model   b) EOMM bridge model

Fig. 5-4 FE model of the short-span bridge

Table 5-2 Major parameters of vehicle (3 axles)

Mass	truck body 1	2612 kg (5746 lbs)
	truck body 2	26113 kg (57448 lbs)
	first axle suspension	490 kg (1078 lbs)
	second axle suspension	808 kg (1777 lbs)
	third axle suspension	653 kg (1436 lbs)
Moment of inertia	Pitching, truck body1	2022 kg.m2 (47882 lbs.ft2)
	Pitching, truck body2	33153 kg.m2 (785083 lbs.ft2)
	Rolling, truck body2	8544 kg.m2 (202327 lbs.ft2)
	Rolling, truck body2	181216 kg.m2 (4291304 lbs.ft2)
Spring stiffness	Upper, 1st axle	242604 N/m (16623 lbs/ft)
	Lower, 1st axle	875082 N/m (59962 lbs/ft)
	Upper, 2nd axle	1903172 N/m (130408 lbs/ft)
	Lower, 2nd axle	3503307 N/m (240052 lbs/ft)
	Upper, 3rd axle	1969034 N/m (134921 lbs/ft)
	Lower, 3rd axle	3507429 N/m (240335 lbs/ft)
Damping coefficient	Upper, 1st axle	2190 N.s/m (150 lbs.s/ft)
	Lower, 1st axle	2000 N.s/m (137 lbs.s/ft)
	Upper, 2nd axle	7882 N.s/m (540 lbs.s/ft)
	Lower, 2nd axle	2000 N.s/m (137 lbs.s/ft)
	Upper, 3rd axle	7182 N.s/m (492 lbs.s/ft)
	Lower, 3rd axle	2000 N.s/m (137 lbs.s/ft)
Length	L1	1.698 m (5.6 ft)
	L2	2.569 m (8.4 ft)
	L3	1.984 m (6.5 ft)
	L4	2.283 m (7.5 ft)
	L5	2.215 m (7.3 ft)
	L6	2.338 m (7.7 ft)
	B	1.1 m (3.6 ft)

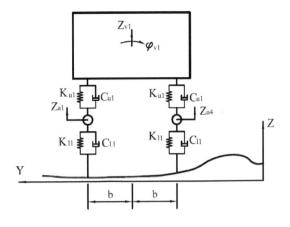

a) Front view

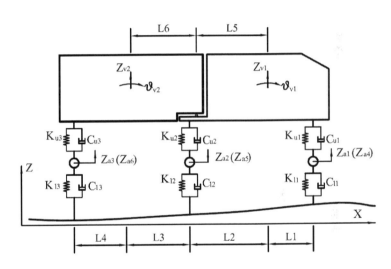

b) Side view

Fig. 5-5 Vehicle model for three axles

107

5.4.2. Modal Analysis

The first several modes of natural frequencies and the mode shapes reflect the bridge's global stiffness and mass distributions, while the high modes reflect the local stiffness and mass distributions. After solving Eq. (5-10), the natural frequencies and mode shapes are obtained via the beam element model, the EOMM shell element model or the real shell element model.

Two finite element models are built for the short span bridge. One is the EOMM shell element model, and the other is the model built with the shell elements with real geometries and material properties. A good match can be found from the comparison of the dynamic properties for the two models. The first ten natural frequencies and the other eight chosen frequencies are listed in Table 5-3. The eight chosen frequencies, which are chosen out of the first 100 modes including some local vibration modes, have more than 1% differences between the two models. Most natural frequencies of the two bridge models have a difference of only less than 1%. Among the first 19 modes, the maximum difference is only 0.46%. However, the mode 20 has a difference of 4%, while modes 97-100 have a difference of 6%. Modes 39, 58 and 59 do have a difference of more than 10% due to the use of the equivalent orthotropic material. In general, the modes match well between the EOMM shell element model and the original shell element model. The results indicate the EOMM shell element model represents a good mass and stiffness distribution of the real structure. Several local vibration modes of the bridge can be obtained via the EOMM shell element model, as well.

Table 5-3 Comparisons of natural frequencies of the short span bridge

No.	Original	EOMM	Dif. (%)	No.	Original	EOMM	Dif. (%)
1	14.826	14.894	0.459	10	15.441	15.485	0.285
2	14.861	14.929	0.453	20	24.401	25.450	4.123
3	14.917	14.982	0.433	39	26.545	39.409	32.64
4	14.986	15.049	0.415	58	41.487	48.405	14.29
5	15.065	15.124	0.387	59	44.151	49.423	10.67
6	15.146	15.202	0.368	97	53.280	56.973	6.482
7	15.227	15.280	0.345	98	53.283	57.243	6.917
8	15.304	15.354	0.326	99	60.351	64.239	6.053
9	15.376	15.423	0.303	100	64.783	69.053	6.183

Due to the complexity of the bridge deck details, only the beam element model (i.e. fish-bone model) and the EOMM shell element model are built for the long-span bridge. Building a finite element model of a long-span bridge with the real configuration would take a great effort. Avoiding such a model is the motivation of the present study, though it would provide a more direct comparison and verification. Six important modes are compared between the two models as shown in Table 5-4. The natural frequencies match well with each other, indicating that the two models have similar stiffness and mass distributions along the bridge in the selected modes. These lower modes are important for the analysis of wind induced vibrations (Cai and Chen 2004).

Based on the comparison results of the short-span and long-span bridges, good precision results can be achieved when the EOMM shell element model is used. Multiple vibration modes for short-span bridges including some local modes and the six main vibration modes for long-span bridges are found to match well with the model built by the original shell element model and the beam element model, respectively. Therefore, the EOMM shell element model is able to model the short-span and long-span bridge with a good precision.

Table 5-4 Dynamic Properties of six main Modes for Donghai Bridge

Mode Number	Frequency (Hz) (Beam element)	Frequency (Hz) (EOMM shell element)	Mode Type
1	0.358	0.374	1st Vertical Mode-Symmetric
2	0.439	0.420	1st Lateral Mode -Symmetric
3	0.511	0.519	1st Vertical Mode - Asymmetric
4	0.590	0.599	1st Torsion Mode - Symmetric
5	1.097	1.149	1st Lateral Mode - Asymmetric
6	1.171	1.151	1st Torsion Mode - Asymmetric

5.4.3. Static Analysis

For the short-span bridge, a series of vertical forces with an equal value of 10,000 N are applied on the nodes at the top right corner of the section. After solving Eq. (5-9), the displacements, strains, and stresses are obtained from both the EOMM shell element model and the shell element model with real geometries and material properties. The three translational displacements and the three rotations in the nodes located at the top right corner of the section are shown in Fig. 5-6. The maximum differences of the three translational displacements are 5% to 15% as shown in Fig. 5-6 (a). In the longitudinal direction U_x, the maximum difference 9.5% is located at ¼ of the span while the maximum differences of 16.1% and 4.7% for the lateral (U_y) and vertical (U_z) directions are observed at the mid-span. The maximum difference of the rotational displacements of ROTY and ROTZ have similar differences of 7% and 11%, respectively while the difference of ROTX is only about 5%. Due to the applied static force, the main deformation for the beam is the torsional rotation (ROTX) and the vertical displacements (U_z). The relatively small differences of the two displacements indicate the present FE modeling scheme simulates correctly on the stiffness terms in the two directions.

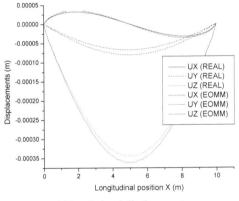

a) Translational displacement

Fig. 5-6 Comparisons of static displacement of short-span bridge

(Fig. 5-6 continued)

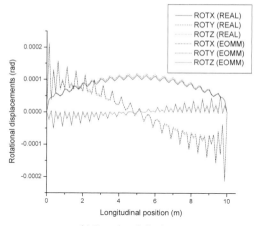

b) Rotational displacement

The strain and stress results of the same nodes are shown in Figs. 5-7 and 5-8. The differences of the normal stress and strain in the longitudinal direction, namely, SX and ε_{xx} are within 10%, in the mid-span of the bridge while the differences of the normal strains and stresses ε_{zz} and SZ are within 2%. Some large differences of strains and stresses in the longitudinal X direction can be found near the end support of the beams due to the restraints set for the longitudinal stiffeners. The shear strain and stress ε_{xy} and SXY have differences of about 10%, and the largest differences are located at the beam support and can reach to 15%. The differences of shear strain and stress ε_{yz} and SYZ are within 5%. The comparisons indicate that the model built with the EOMM method can predict the displacements with a maximum difference of 12% and the strains and stresses with a maximum difference of 10%, except for the end supports. Therefore, the EOMM method can be used for static analysis of steel bridges, and reasonable results of displacements, strains and stresses can be achieved.

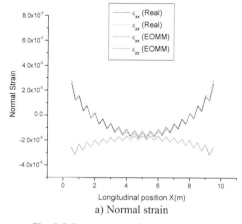

a) Normal strain

Fig. 5-7 Comparisons of static strain for short-span bridge

110

(Fig. 5-7 continued)

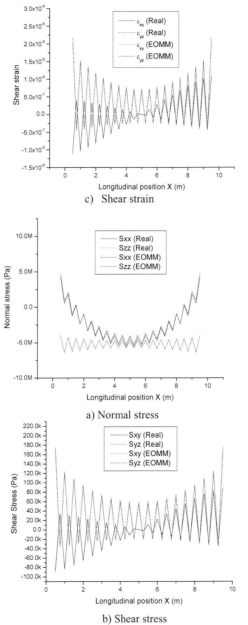

c) Shear strain

a) Normal stress

b) Shear stress

Fig. 5-8 Comparisons of static stress for short-span bridge

5.4.4. Vehicle-Bridge Dynamic Analysis

For a demonstrational purpose, the dynamic response of the short span bridge due to the passing of an HS20-44 truck with a speed of 60m/s is obtained. A road profile with an average road roughness classification index (ISO 1995) is generated as the excitation input of the dynamic system through an inverse Fourier transformation based on the power spectral density function listed in Eq. (5-22). After solving the equations of motions by the Rouge-Kutta method in the time domain, the time history of the displacements and stresses are obtained.

In order to validate the efficiency of the EOMM shell element in modeling bridges, the first 100 modes are used to predict the dynamic responses from the vehicle-bridge dynamic analysis. Meanwhile, the modes with large differences are intentionally kept in the dynamic analysis using mode superposition techniques. The three translational and rotational displacements at the top right corner of the section in the mid-span match well between the original shell element model and the EOMM shell element model as shown in Fig. 5-9.

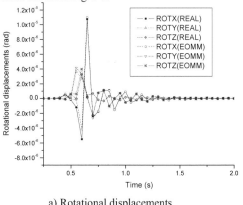

a) Rotational displacements

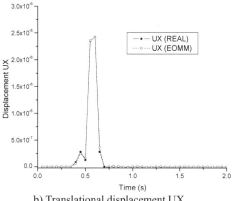

b) Translational displacement UX

Fig. 5-9 Comparisons of dynamic displacemet history (100modes) for short-span bridge

112

(Fig. 5-9 continued)

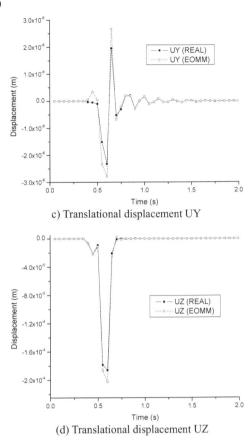

c) Translational displacement UY

(d) Translational displacement UZ

The rotational displacements from the two vehicle-bridge dynamic systems with different modeling methods match well with each other as shown in Fig. 5-9 (a). In the longitudinal direction, the displacements obtained from the two vehicle-bridge dynamic systems have a difference of less than 1%. However, in the lateral (Uy) and vertical (Uz) directions, the EOMM shell element model over-predicts the maximum lateral displacement at the time of 0.6s to a ratio of 1.2 as shown in Fig. 5-9(c)-(d). Based on Eq. (5-20), the stresses can be obtained as shown in Fig. 5-10. For the three main stresses, including the normal stress S_{xx}, S_{zz} and shear stress S_{xz}, they have the same trend when the vehicle is moving along the bridge. The differences for the maximum stress S_{xx}, S_{zz} and S_{xz} are 11.3%, -4.83% and -1.64%, respectively. On an equivalent fatigue damage basis, the revised equivalent stress ranges for the three stresses can be obtained via Eq. (5-21) and listed in Table 5-5. Even though the stresses in some directions vary between the original FE model and the EOMM model, including normal stress S_{xx}, S_{zz} and shear stress S_{xz}, the revised equivalent stress ranges only vary less than 11%. The differences for normal stress S_{zz} is 3%, while the differences for normal stress S_{xx} and shear stress S_{xz} are about 10%.

113

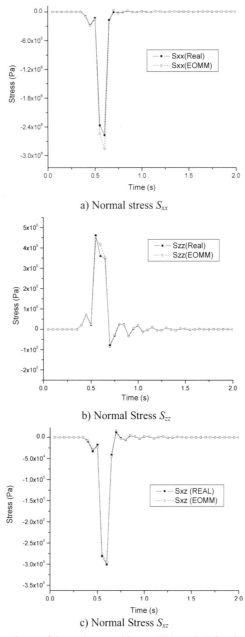

a) Normal stress S_{xx}

b) Normal Stress S_{zz}

c) Normal Stress S_{xz}

Fig. 5-10　　Comparisons of dynamic stress history (100 modes) for short-span bridge

Table 5-5 Comparisons of revised equivalent stress ranges for short span bridge

Category	Original		EOMM		Difference	
	100 modes	19 modes	100 modes	19 modes	100 modes	19 modes
S_{xx}	2.58 Mpa	13.5 Kpa	2.87 Mpa	14.0 Kpa	11.5%	4.3%
S_{zz}	0.50 Mpa	12.3 Kpa	0.49 Mpa	12.9 Kpa	3.3%	5.1%
S_{xz}	0.31 Mpa	2.72 Kpa	0.30 Mpa	2.45 Kpa	0.9%	-9.9%

The dynamic effects on long-span bridges from wind loads are dominated by the low frequency vibrations (Cai and Chen 2004). Only several basic vibration modes contribute most to the dynamic responses, such as the listed modes for Donghai Bridge in Table 5-4. For the present numerical example, the stress differences arise from the differed mode shapes and natural frequencies. Due to the simplification of the bridge details with equivalent orthotropic materials, the EOMM model cannot achieve the same for every mode shape and natural frequency. The dynamic stresses from the EOMM shell element model and the original shell element model are obtained based on the first 19 modes that have less than 1% differences in natural frequencies. The results are compared and shown in Fig. 5-11. All of the two normal stresses and one shear stress have a good match between the results from the EOMM shell element model and the original shell element model. Based on the equivalent fatigue damage, the revised equivalent stress ranges for the three stresses can also be obtained via Eq. (5-21) and the results are listed in Table 5-5. The differences of the normal stress S_{xx} drops to 4%. However, the differences for normal stress S_{zz} and shear stress S_{xz} increase to 5% and 9.9%, respectively. For the engineering purpose of fatigue analysis, the EOMM modeling scheme is effective for estimating stresses for fatigue reliability analysis and such differences are acceptable.

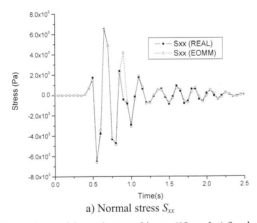

a) Normal stress S_{xx}

Fig. 5-11   Comparisons of dynamic stress history (19 modes) for short-span bridge

(Fig. 5-11 continued)

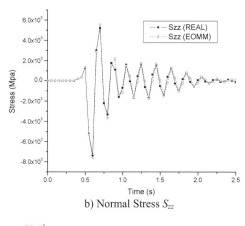

b) Normal Stress S_{zz}

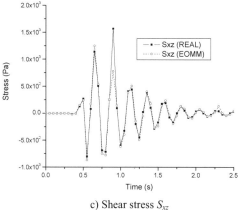

c) Shear stress S_{xz}

It is noteworthy that sufficient modes for modal analysis are needed to calculate the stress in bridge details. The obtained stresses are different between the results from the dynamic system with superposition of 19 modes and 100 modes. Different modes might have a different weight in different directional stresses. Therefore, bridges of varying types and span lengths will differ with each other greatly in terms of required modes and need to be carefully considered before carrying out the dynamic analysis using mode superposition techniques. The discussion on this subject is, however, beyond the scope of this study.

5.5 Concluding Remarks

An effective FE model is important to evaluate the structural performance under multi-scale dynamic loads, for instance the wind loads in a kilo-meter scale and the vehicle loads in a meter scale. The superposition of the stresses from the multi-scale dynamic loads might cause serious fatigue damage accumulation for long-span bridges. This paper presents a multiple scale modeling and simulation scheme based on the EOMM method. Bridge deck plates with multiple stiffeners are modeled as the elements using an equivalent orthotropic material and geometry. The bridges are assembled with simplified equivalent shell elements with the same position as the original shell element. Based on the comparison of results from modal, static and vehicle-bridge dynamic analyses,

116

the following conclusions are drawn: The EOMM method can be used to build the FE model with good precision in vibration modes including the main vibration modes and several local vibration modes; The model built by the EOMM has good precision in predicting static displacements, strains and stresses; The dynamic stresses from the model built by the EOMM have good precision if only the matched modes are used for the mode superposition techniques in the dynamic analysis.

Based on the multiple scales modeling scheme, it is possible to predict a reasonable static and dynamic response of the bridge details since the EOMM model is capable of including the global vibration modes and local vibration modes of the original model with refined structural details. Due to the approximation and assumption of some material properties, some discrepancies can be found in some directions of the dynamic stresses if different mode shapes are included in the mode superposition procedure. However, if only the matched modes are included, the differences of dynamic stress from the EOMM model and the original model drops from 130% to 20%, which is acceptable for an engineering approach to predict fatigue damage accumulations. Therefore, based on the EOMM model, it is possible to calculate the dynamic effects in multiple scales, namely, from the wind loads in a kilo-meter scale in a low frequency region if enough global vibration modes are included and the vehicle loads in a meter scale in a high frequency region if enough local modes are included in the analysis.

5.6 References

Batoz, J. L., Bathe, K. J., and Ho, L. W. (1980). "A Study of Three-Noded Triangular Plate Bending Elements." *International Journal for Numerical Methods in Engineering*, 15, 1771-1812.

Cai, C. S., and Chen, S. R. (2004). "Framework of vehicle-bridge-wind dynamic analysis." *Journal of Wind Engineering and Industrial Aerodynamics*, 92(7-8), 579-607.

Chan, T. H. T., Yu, Y., Wong, K. Y., & Li, Z. X. (2008). Condition-assessment-based finite element modeling of long-span bridge by mixed dimensional coupling method. (J. Gao, J. Lee, J. Ni, L. Ma, & J. Mathew, Eds.)Mechanics of Materials. Springer. Retrieved from http://eprints.qut.edu.au/16720/

Chan, T. H. T., Zhou, T. Q., Li, Z. X., and Guo, L. (2005). "Hot spot stress approach for Tsing Ma Bridge fatigue evaluation under traffic using finite elment method." *Structural Engineering and Mechanics*, 19(3), 261-79.

Deng, L., and Cai, C. S. (2010). "Development of dynamic impact factor for performance evaluation of existing multi-girder concrete bridges." *Engineering Structures*, 32(1), 21-31.

Dodds, C. J., and Robson, J. D. (1973). "The Description of Road Surface Roughness." *Journal of Sound and Vibration*, 31(2), 175-183.

Du, B. (2006). "Dynamic Characteristics of Suspension Bridges with Twin-Box Girder Considering Nonlinearity," Ph.D Thesis, Tongji University, Shanghai.

Gu, M., Xu, Y. L., Chen, L. Z., and Xiang, H. F. (1999). "Fatigue life estimation of steel girder of Yangpu cable-stayed Bridge due to buffeting." *Journal of Wind Engineering and Industrial Aerodynamics*, 80(3), 383-400.

Honda, H., Kajikawa, Y., and Kobori, T. (1982). "Spectra of Road Surface Roughness on Bridges." *Journal of the Structural Division*, 108(ST-9), 1956-1966.

International Standard Organization. (1995). "Mechanical vibration - Road surface profiles - Reporting of measured data." Geneva.

Kansara, K. (2004). "Development of Membrane, Plate and Flat Shell Elements in Java," Master Thesis, Virginia Polytechnic Institue & State University, Blacksburg, Virginia.

Keating, P. B., and Fisher, J. W. (1986). "Evaluation of Fatigue Tests and Design Criteria on Welded Details." NCHRP Report 286, Transportation Research Board, Washington, D.C.

Larsen, A., Savage, M., Lafreniere, A., Hui, M. C. H., and Larsen, S. V. (2008). "Investigation of vortex response of a twin box bridge section at high and low Reynolds numbers." *Journal of Wind Engineering and Industrial Aerodynamics*, 96(6-7), 934-944.

Li, Z. X., Zhou, T. Q., Chan, T. H. T., and Yu, Y. (2007). "Multi-scale numerical analysis on dynamic response and local damage in long-span bridges." *Engineering Structures*, 29(7), 1507-1524.

Mises, R. v. (1945). "On Saint Venant's principle." *Bulletin of the American Mathematical Society*, 51(3), 555-562.

Monaghan, D. J. (2000). "Automatically coupling elements of dissimilar dimension in finite element analysis," Ph.D Thesis, The Queen's University of Belfast.

Shi, X., Cai, C. S., and Chen, S. (2008). "Vehicle Induced Dynamic Behavior of Short-Span Slab Bridges Considering Effect of Approach Slab Condition." *Journal of Bridge Engineering*, 13(1), 83-92.

Troitsky, M. S. (1987). *Orthotropic Bridges - Theory and Design, 2nd ed.*, The James F. Lincoln Arc Welding Foundation, Clevaland, OH.

Wang, T.-L., and Huang, D. (1992). "Computer modeling analysis in bridge evaluation." Florida Department of Transportation, Tallahassee, FL.

Wolchuk, R. (1963). Design Manual for Orthotropic Steel Plate Deck Bridges. Chicago, IL: American Institute of Steel Construction.

Wu, C., Zeng, M.G. and Dong, B. (2003) "Report on the Performance of Steel-Concrete Composite Beam of Donghai Cable-stayed Bridge." Department of Bridge Engineering, Tongji University, Shanghai.

Xu, Y. L., Liu, T. T., and Zhang, W. S. (2009). "Buffeting-induced fatigue damage assessment of a long suspension bridge." International Journal of Fatigue, 31(3), 575-586.

Zhang, W., C.S. Cai. (2011). "Fatigue Reliability Assessment for Existing Bridges Considering Vehicle and Road Surface Conditions", Journal of Bridge Engineering, doi:10.1061/ (ASCE) BE. 1943-5592.0000272

Zhang, Z. T., and Ge, Y. J. (2003). "Static and Dynamic Analysis of Suspension Bridges Based on Orthotropic Shell Finite Element Method." Structural Engineers, 2003(4).

CHAPTER 6   FATIGUE RELIABILITY ASSESSMENT FOR LONG-SPAN
BRIDGES UNDER COMBINED DYNAMIC LOADS FROM WINDS
AND VEHICLES

6.1 Introduction

Fatigue is one of the main forms of structural damages and failure modes caused by repeated dynamic load effects, for instance, wind loads and vehicle loads. With the great increase of span lengths, bridges are becoming more flexible and more vulnerable to wind induced vibrations. Virlogeux (1992) and Gu et al. (1999), by neglecting the vehicle effects, conducted buffeting induced fatigue analysis on two cable-stayed bridges and the fatigue life was found to be much longer than the design life of the bridges. Based on the recorded data of the Tsing Ma Bridge, Xu et al. (1999) found that the monsoon wind-induced fatigue damage is not significant. In addition, many works have been carried out on the vehicle-bridge dynamic analysis or dynamic analysis of long-span bridges under wind, railway and highway loadings (Byers et al. 1997a, b, Guo and Xu 2001, Cai and Chen 2004, Xu et al. 2009, Chen and Wu 2010, Chen et al. 2011 a, b). However, most of these studies focus on dynamic displacements and accelerations by using a simple finite element model of the bridge without including structural details (Chen et al 2011a). Systematic approaches on the fatigue reliability assessment of long-span bridges are still lacking, which can consider bridge's structural details, for instance, the stiffeners installed on the orthotropic bridge decks, and multiple random variables, for instance, vehicle speed, vehicle type, and wind velocity and direction. Since long-span bridges often serve as the backbones of main transportation lines to support daily operation and hurricane evacuations, the structural reliability should be carefully assessed and predicted especially for the superposed multiple dynamic loads to ensure the structural safety.

To make an accurate estimation of the fatigue life of existing bridges, it is necessary to predict a reasonable future stress range history due to various traffic loadings under various wind environments and road surface conditions. Such data can be obtained either from on-site strain measurements (Chan et al. 2001, Kwon and Frangopol 2010) or structural dynamic analysis of bridges. However, stress range spectra for bridges are strongly site-specific due to different vehicle types and speed distributions, road roughness conditions, and bridge types (Laman and Nowak 1996). Instead, numerical simulations can be used in a more versatile way to simulate complex scenarios including varied wind velocities, vehicle speeds, road roughness conditions, vehicle types, and driver operation characteristics (Chen and Wu 2010). In the past decades, there have been a number of studies on the vehicle-bridge-wind dynamic system for dynamic stress analysis of long span bridges (Guo and Xu 2001, Cai and Chen 2004, Chen and Wu 2010, Chen et al. 2011b).

In the dynamic system, interactions between the bridge and vehicles are modeled as coupling forces between the tires and the road surface. The coupling forces were proven to be significantly affected by the vehicle speed and road roughness conditions and resulted in significant effects on the dynamic responses of short span bridges (Shi et al. 2008, Deng and Cai 2010, Zhang and Cai 2011). Differently, the dynamic effect for a long-span suspension bridge from vehicles can be neglected and a simplified engineering approach was proven to be effective based on the fatigue analysis of Tsing Ma suspension bridge with a main span length of 1377m (4518ft) (Chen et al. 2011b). However, for different types of long-span bridges with the span length ranging from a few hundreds to thousands meters, a more general framework to ensure the structural safety is essential. Specifically, the framework should include more effective and accurate modeling methods beyond a simple beam element model, more reasonable procedures to generate dynamic stress histories for multiple traffic or wind conditions and more comprehensive fatigue probability analysis scheme that could include progressive fatigue damage accumulation in a bridge's life cycle. In addition, the framework should have the capacity to include multiple random variables for the dynamic loads in a bridge's life cycle for the vehicle-bridge-wind dynamic system, for instance, road profiles, vehicle speeds, and wind velocities and directions, etc. Certain retrofitting actions can be performed based on the results from the fatigue reliability analysis, for instance, repairing the structure, replacing the structure or changing the operation of the structure (Byers et al. 1997 a).

In the present study, a general framework of fatigue reliability assessment for long-span bridges under combined dynamic loads from winds and vehicles is proposed and summarized as the following procedures. First of all, a computationally efficient modeling scheme is needed to build an accurate finite element model with the possibility to acquire dynamic stresses in bridge's details. The equivalent orthotropic material modeling (EOMM) scheme is used in the present study for modeling long-span bridges with complicated structural details such as the longitudinal stiffeners in the girder using simplified shell elements with equivalent material properties. This modeling scheme is proven to be effective in static and dynamic analysis via a case study for the bridges with complicated structural details in a preliminary study. Therefore, the calculation cost can be saved and accuracy is preserved. In step 2, the dynamic stress histories for given structural details are obtained for a given vehicle speed, wind velocity and direction, and road roughness condition by solving the equations of motions of the vehicle-bridge-wind dynamic system. In step 3, the random variables for the vehicle-bridge-wind dynamic system for each block of stress cycles need to be specified. The size of the stress cycle blocks might vary with each other, for instance, the stress cycle block is assumed lasting for one hour for wind loads and one day for vehicle loads. The road roughness condition needs to be specified from either the current condition assessment or the prediction based on the traffic information. The vehicle speed and wind velocity and direction are generated based on the traffic or meteorological data. Therefore, the dynamic stress histories in each block of stress cycles are randomly generated based on the results from the last step. In step 4, the fatigue damage accumulation rule is specified, for instance, as being a linear fatigue damage model or a nonlinear continuous fatigue damage model. After counting the number of stress cycles at different stress range levels using rainflow counting method, a fatigue damage increment ΔD_i can be obtained using the fatigue damage accumulation rule. In step 5, a failure function or limit state function (LSF) need to be specified. The probability of failure for the fatigue damage D_i at the end of each block of stress cycles and the cumulative probability of failure can be obtained. The calculations go back to step 3 for the next block of stress cycles if the cumulative probability of failure has not increased to the target value corresponding to the target reliability index. Therefore, the progressive fatigue damage accumulation in the bridge's life cycle is calculated and the fatigue life and reliability for the given structure details in a bridge's life cycle is obtained.

The paper is organized as the following five main sections. In the first section, the equivalent orthotropic modeling scheme for long-span bridges is introduced and the prototype bridge is introduced. In the second section, the vehicle-bridge-wind dynamic system and its parameters are introduced. The equations of motions, the modeling of the dynamic loads, the load distributions on nodes, and the principles for generating stochastic random road profiles are detailed. In the third section, the stress cycle blocks are defined and their acquisition methods are introduced with traffic simulation and wind environment information included. In the fourth section, the fatigue damage model and limit state function is defined. In the last section, selected results are provided to assess the fatigue reliability of long-span bridges from wind loads, vehicle loads, and their combined loads.

6.2 Finite Element Modeling Scheme for Long-Span Bridges

6.2.1. Orthotropic Bridge Deck

Most of the metallic alloys and thermoset polymers are considered isotropic, whose properties are independent of directions. In their stiffness and compliance matrices, only two elastic constants, namely, the Young's modulus E and the Poisson's ratio v are independent. In contrast, the orthogonal materials have independent material properties in at least two orthogonal planes of symmetry. A total number of 21 elastic constants are needed for fully anisotropic materials without any plane of symmetry.

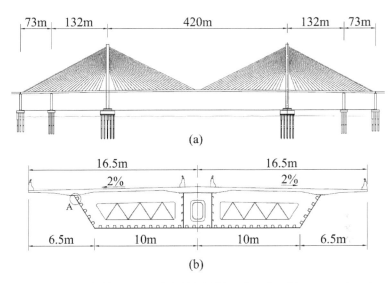

Fig. 6-1 Prototype bridge (Wu et al. 2003): (a) Elevation view (b) cross-section view

In order to enhance the bending resistance of the steel plate to carry local loads from vehicle wheels, orthotropic bridge decks were developed by German engineers in the 1950s (Wolchuk 1963). As a result, the total cross-sectional area of steel in the plate was increased and the overall bending capacity of the deck and the resistance of the plate to buckling were increased, as well. The creative orthotropic bridge design not only offered excellent structural characteristics, but was also economical to build (Troisky 1987). From short span bridges to long span cable-supported bridges, the orthotropic bridge designs have been used throughout the world, for instance, the Golden Gate Bridge and the Akashi-Kaikyo Bridge in Japan. In addition to the bridge deck, the orthotropic steel plates have been used in the other parts of the bridge deck systems, such as the cross plates or the side plates. For example, the Donghai Cable-stayed Bridge in China has a main span of 420m. It has a prestressed concrete deck, while the web, cross plates and bottom plates have various open and closed rib stiffeners, as shown in Fig. 6-1 and will discussed later (Wu et al 2003). In order to model bridges with small stiffeners, large computational efforts are needed if all the stiffeners are modeled in details and it is almost impossible to carry out the dynamic analysis on such a model. Due to the orthotropic properties of the deck plate, it is possible to use an equivalent orthotropic shell element to model the plate with various stiffeners.

6.2.2. Equivalent Orthotropic Material Modeling Method

In the longitudinal and lateral directions, the multiple open or closed ribs provide varied stiffness to the steel plates in the bridge deck, such as the longitudinal trapezoidal stiffeners shown in Fig. 6-1. In order to avoid an unmanageably large number of elements and degree of freedoms involved in solving the equations of the motions of the bridge, an equivalent orthotropic shell

121

element was used. In the present study, the following equivalent rules for the equivalent orthotropic shell element are applied.

Firstly, the equivalent orthotropic shell element has the same bending stiffness in a unit width as the original configuration:

$$E_x \frac{d^3}{12} = EI_x, E_y \frac{d^3}{12} = EI_y \tag{6-1}$$

where, d is the equivalent thickness of the shell, E_x and E_y are the equivalent elastic modulus in two orthogonal x and y directions, I_x and I_y are the moment of inertia in x and y directions, and E is the elastic modulus of the original material. The x direction is along the bridge and y direction is perpendicular to x direction in the shell plane.

Secondly, the equivalent orthotropic shell has the same longitudinal axial stiffness and shear stiffness as the original configuration:

$$E_x d = EA, G_{xy} d = Gt \tag{6-2}$$

where, A is the area of the unit width of the shell, G_{xy} is the equivalent shear modulus, t is the thickness of the plate, and G is the shear modulus of the original material.

Thirdly, the equivalent orthotropic shell element has the same weight as the original element:

$$\rho_e d = \rho A \tag{6-3}$$

where, ρ_e is the equivalent density of the shell element and ρ is the original density of the structure material.

Based on the equivalent rules, the material properties of the equivalent shell element can be obtained from Eqs. (6-1) to (6-3):

$$d = \sqrt{\frac{12I_x}{A}}, E_x = EA\sqrt{\frac{A}{12I_x}}, E_y = \frac{12EI_y}{d^3}A, G_{xy} = \frac{Gt}{d}, \rho_e = \frac{\rho A}{d} \tag{6-4}$$

In addition, the locations of each equivalent shell element remain unchanged and the following assumptions are made in order to define the material matrix. The Poisson's ratio v_{ij} are all assumed to be zeros and E_z, G_{yz} and G_{xz} are given a relatively small value compared with other modulus in other directions. The material matrix for the equivalent shell element is listed in Eq. (6-5) as:

$$[E]^{eq} = \begin{bmatrix} E\sqrt{\frac{A^3}{12I_x}} & 0 & 0 & 0 & 0 & 0 \\ 0 & \frac{12EI_y}{d^3} & 0 & 0 & 0 & 0 \\ 0 & 0 & E_z & 0 & 0 & 0 \\ 0 & 0 & 0 & 2G_{yz} & 0 & 0 \\ 0 & 0 & 0 & 0 & 2G_{zx} & 0 \\ 0 & 0 & 0 & 0 & 0 & 2\frac{Gt}{d} \end{bmatrix} \tag{6-5}$$

Based on the equivalent orthotropic material modeling (EOMM) method, the multi-scale modeling scheme can be used to model long span bridges. Bridge details with multiple complicated stiffeners are modeled as the equivalent shell element using equivalent orthotropic material and geometry. The material matrix for the equivalent orthotropic material is obtained and the element stiffness matrix is changed accordingly. As a result, the coupled effects between different directions in the material matrices disappear. Since the equivalence is only based on its longitudinal stiffness EI_x, lateral stiffness EI_y, and shear stiffness $G_{xy}t$, differences are expected for the output results related to the other elastic and shear modulus of E_z, G_{yz}, and G_{zx}. Compared with the simple beam element model, which is also called "fish-bone" FE model, the EOMM model has a better modeling of the stiffness and mass distribution of the bridge deck sections.

6.2.3.   Scheme Validation and the Prototype Bridge

To demonstrate the EOMM method and validate its efficiency in predicting the static and dynamic responses of the bridge, a simplified short span beam bridge and a long span bridge are analyzed in a preliminary study. Bridge deck plates with multiple stiffeners are modeled as the equivalent shell element using equivalent orthotropic material and geometry with the same longitudinal and lateral stiffness in a unit width and shear stiffness in the flat shell plane. Based on the modeling scheme, it is possible to predict a reasonable static and dynamic response of the bridge details since the EOMM model is capable of including the refined structural details. The static and dynamic response and dynamic properties of a simplified short span bridge from the EOMM model are obtained. The results match well with those obtained from the original model with real geometry and materials. The EOMM model for a long-span cable-stayed bridge is built with good precision on dynamic properties, which can be used for the wind induced fatigue analysis. Based on the modeling scheme, it is possible to predict the multi-scale dynamic loads' effects, for instance, the wind induced vibrations of low frequency in kilo-meter scale and the vehicle induced vibrations of high frequency in meter scale.

In the present study, the Donghai Cable-stayed Bridge is used to serve as the prototype bridge. It has a main span of 420m and is located in a typhoon zone of east China. The deck of the girder is made of prestressed concrete, while the web, cross plates, and bottom plates with multiple various stiffeners are made of steel (Ge et al. 2003, Wu et al. 2003). In order to obtain the stress histories in the bridge details, equivalent orthogonal shell elements are used to model the complicated bridge deck plate with multiple open or closed ribs, such as longitudinal trapezoidal stiffeners as shown in Fig. 6-1. Therefore, it is possible to predict the dynamic response of the bridge detail and the effects from distortion, constrained torsion, and shear lag can be taken into account.

Due to the complexity of the bridge deck details, only the beam element model (i.e. fish-bone model) and the EOMM shell element model are built for the long-span bridge. Building a finite element model of long-span bridges with the real configuration would take a great effort. Avoiding such a model is the motivation of the present study, though it would provide a more direct comparison and verification. Six important modes are compared between the two models as shown in Table 1. The differences between the two models are relatively small and less than 5%. The well match of natural frequencies indicates that the two models have similar stiffness and mass distributions along the bridge in the selected modes. It is noteworthy that these lower modes are important for the analysis of wind induced vibrations (Cai and Chen 2004).

Based on the static analysis, the present study focuses on the fatigue analysis at detail A as shown in Fig. 6-1, which suffers from large stress ranges due to the passage of vehicles. Both the membrane stress and bending stress are included in the S_y stress, which is an in-plane stress and along the web.

Table 6-1 Natural frequencies of the six modes for Donghai Bridge

Mode Number	Frequency (Hz) (Beam element)	Frequency (Hz) (Orthogonal shell element)	Mode Type
1	0.358	0.374	1st Vertical Mode-Symmetric
2	0.439	0.420	1st Lateral Mode -Symmetric
3	0.511	0.519	1st Vertical Mode - Asymmetric
4	0.590	0.599	1st Torsion Mode - Symmetric
5	1.097	1.149	1st Lateral Mode - Asymmetric
6	1.171	1.151	1st Torsion Mode - Asymmetric

6.3 Vehicle-Bridge-Wind Dynamic System

6.3.1. Equations of Motion for Vehicle-Bridge-Wind System

In the present study, the vehicle is modeled as a combination of several rigid bodies connected by several axle mass blocks, springs, and damping devices (Cai and Chen 2004). The tires and suspension systems are idealized as linear elastic spring elements and dashpots. The contact between the bridge deck and the moving tire is assumed to be a point contact. The model can be used to simulate vehicles on highway roads or bridges with axle number varying from two to five. The bridge can be modeled by using different types of elements such as beam element, shell element, and solid element, depending on the bridge type. The mass matrix and stiffness matrix can be obtained by the conventional finite element method. The motions of the bridge and the vehicle can be expressed as the following equations:

$$[M_b]\{\ddot{d}_b\}+[C_b]\{\dot{d}_b\}+[K_b]\{d_b\}=\{F_b^c\}+\{F_b^w\} \tag{6-6}$$

$$[M_v]\{\ddot{d}_v\}+[C_v]\{\dot{d}_v\}+[K_v]\{d_v\}=\{F_v^G\}+\{F_v^c\}+\{F_v^w\} \tag{6-7}$$

where, [M] are the mass matrices, [C] are the damping matrices and [K] are the stiffness matrices; $\{F_b^c\}$ is wheel-bridge contact forces on bridge, $\{F_b^w\}$ is the vector of wind effects on the bridge, $\{F_v^G\}$ is the self-weight of vehicle, $\{F_v^c\}$ is the vector of wheel-road contact forces acting on the vehicle, and $\{F_v^w\}$ is the vector of wind effects on the vehicle. The two equations are coupled through the contact condition, i.e., the interaction forces $\{F_v^c\}$ and $\{F_b^c\}$, which are action and reaction forces existing at the contact points of the two systems and can be stated as a function of deformation of the vehicle's lower spring:

$$\{F_b^c\}=-\{F_v^c\}=[K_l]\{\Delta_l\}+[C_l]\{\dot{\Delta}_l\} \tag{6-8}$$

where $[K_l]$ and $[C_l]$ are the coefficients of the vehicle's lower spring and damper; and Δ_l is the deformation of the lower springs of the vehicle. The relationships among the vehicle-axle-suspension displacement Z_a, displacement of the bridge at wheel-road contact points Z_b, deformation of the lower springs of the vehicle Δ_l, and road surface profile $r(x)$ are:

$$Z_a = Z_b + r(x) + \Delta_l \tag{6-9}$$

$$\dot{Z}_a = \dot{Z}_b + \dot{r}(x) + \dot{\Delta}_l \tag{6-10}$$

where $\dot{r}(x)=(dr(x)/dx)\cdot(dx/dt)=(dr(x)/dx)\cdot V(t)$ and $V(t)$ is the vehicle velocity.

Therefore, the contact force $\{F_v^c\}$ and $\{F_b^c\}$ between the vehicle and the bridge are derived as:

$$\{F_b^c\} = -\{F_v^c\} = [K_l]\{Z_a - Z_b - r(x)\} + [C_l]\{\dot{Z}_a - \dot{Z}_b - \dot{r}(x)\}$$ (6-11)

6.3.2. Modeling of Road Surface Roughness

Based on the studies carried out by Dodds and Robson (1973) and Honda et al. (1982), the long undulations in the roadway pavement could be assumed as a zero-mean stationary Gaussian random process and it could be generated through an inverse Fourier transformation (Wang and Huang 1992):

$$r(x) = \sum_{k=1}^{N} \sqrt{2\phi(n_k)\Delta n} \cos(2\pi n_k x + \theta_k)$$ (6-12)

where θ_k is the random phase angle uniformly distributed from 0 to 2π; $\phi()$ is the power spectral density (PSD) function (m³/cycle) for the road surface elevation; and n_k is the wave number (cycle/m). The PSD functions for road surface roughness were developed by Dodds and Robson (1973), and three groups of road classes were defined with the values of roughness exponents ranging from 1.36 to 2.28 for motorways, principal roads, and minor roads. In order to simplify the description of road surface roughness, both of the two roughness exponents were assumed to have a value of two and the PSD function was simplified by Wang and Huang (1992) as:

$$\phi(n) = \phi(n_0)(\frac{n}{n_0})^{-2}$$ (6-13)

where $\phi(n)$ is the PSD function (m³/cycle) for the road surface elevation; n is the spatial frequency (cycle/m); n_0 is the discontinuity frequency of $1/2\pi$ (cycle/m); and $\phi(n_0)$ is the road roughness coefficient (m³/cycle) whose value is chosen depending on the road condition.

6.3.3. Modeling of Wind Force Vectors

The external wind loading on the dynamic system consists of wind loading on bridges and simplified quasi-steady wind forces on vehicles (Chen and Cai 2004). For bridges immersing in the wind, the total wind forces on the center of bridge elasticity $\{F_b^w\}$ in Eq. (6-6) can be expressed as:

$$\{F_b^w\} = \begin{bmatrix} 0 \\ L_b^w(x,t) \\ D_b^w(x,t) \\ M_b^w(x,t) \\ 0 \\ 0 \end{bmatrix} = \begin{bmatrix} 0 \\ L_{st} + L_{ae}(x,t) + L_b(x,t) \\ D_{st} + D_{ae}(x,t) + D_b(x,t) \\ M_{st} + M_{ae}(x,t) + M_b(x,t) \\ 0 \\ 0 \end{bmatrix}$$ (6-14)

where the subscripts "st", "ae" and "b" refer to the static, self-exited, and buffeting force component due to wind, respectively.

The static wind force of unit span length on the center of bridge elasticity can be expressed as:

$$L_{st} = \frac{1}{2}\rho U^2 B \cdot C_L; \quad D_{st} = \frac{1}{2}\rho U^2 B \cdot C_D; \quad M_{st} = \frac{1}{2}\rho U^2 B^2 \cdot C_M$$ (6-15)

where ρ is the air density; U is the mean wind velocity on the elevation of the bridge; B is the bridge

125

deck width; and C_L, C_D and C_M are the lift, drag, and moment static wind force coefficients for the bridges, respectively, that are usually obtained from section model wind tunnel tests of the bridge deck.

The self-excited force on the center of bridge elasticity can be expressed as (Chen and Cai 2004):

$$\begin{cases} L_{se} = \dfrac{1}{2}\rho U^2 B\left[KH_1^* \dfrac{\dot{h}(x,t)}{U} + KH_2^* \dfrac{B\dot{\alpha}(x,t)}{U} + K^2 H_3^* \alpha(x,t) + K^2 H_4^* \dfrac{h(x,t)}{B} + KH_5^* \dfrac{\dot{p}(x,t)}{U} + K^2 H_6^* \dfrac{p(x,t)}{B} \right] \\[4mm] D_{se} = \dfrac{1}{2}\rho U^2 B\left[KP_1^* \dfrac{\dot{p}(x,t)}{U} + KP_2^* \dfrac{B\dot{\alpha}(x,t)}{U} + K^2 P_3^* \alpha(x,t) + K^2 P_4^* \dfrac{p(x,t)}{B} + KP_5^* \dfrac{\dot{h}(x,t)}{U} + K^2 P_6^* \dfrac{h(x,t)}{B} \right] \\[4mm] M_{se} = \dfrac{1}{2}\rho U^2 B^2\left[KA_1^* \dfrac{\dot{h}(x,t)}{U} + KA_2^* \dfrac{B\dot{\alpha}(x,t)}{U} + K^2 A_3^* \alpha(x,t) + K^2 A_4^* \dfrac{h(x,t)}{B} + KA_5^* \dfrac{\dot{p}(x,t)}{U} + K^2 A_6^* \dfrac{p(x,t)}{B} \right] \end{cases} \quad (6\text{-}16)$$

where $K = B\omega / U$ is the reduced frequency; H_i^*, P_i^* and A_i^* (i=1 to 6) are the flutter derivatives of the bridge obtained from the wind tunnel tests of the bridge deck; ω is the vibration frequency of the system; and the dot on the cap denotes the derivative with respect to the time.

The buffeting forces for a unit span in vertical, lateral, and torsional directions on the center of bridge elasticity are (Chen and Cai 2004):

$$\begin{cases} L_b = \dfrac{1}{2}\rho U^2 B\left[C_L\left(2\dfrac{u(t)}{U}\right) + (C_L' + C_D)\dfrac{w(t)}{U} \right] \\[4mm] D_b = \dfrac{1}{2}\rho U^2 B\left[C_D\left(2\dfrac{u(t)}{U}\right) + (C_D' - C_L)\dfrac{w(t)}{U} \right] \\[4mm] M_b = \dfrac{1}{2}\rho U^2 B^2\left[C_M\left(2\dfrac{u(t)}{U}\right) + C_M'\dfrac{w(t)}{U} \right] \end{cases} \quad (6\text{-}17)$$

where $u(t)$ and $w(t)$ are the horizontal and vertical components of wind turbulent velocity, respectively; and the prime denotes the derivative with respect to the attack angle of wind.

Wind action on a running vehicle includes static and dynamic load effects. The quasi-static wind forces on vehicles are adopted since a transient type of force model is not available (Baker 1994, Chen and Cai 2004):

$$\begin{cases} F_{wx} = \dfrac{1}{2}\rho U_r^2 C_D(\psi)A; \quad F_{wy} = \dfrac{1}{2}\rho U_r^2 C_S(\psi)A; \quad F_{wz} = \dfrac{1}{2}\rho U_r^2 C_L(\psi)A \\[4mm] M_{w\phi} = \dfrac{1}{2}\rho U_r^2 C_R(\psi)Ah_v; \quad M_{w\theta} = \dfrac{1}{2}\rho U_r^2 C_P(\psi)Ah_v; \quad M_{wz} = \dfrac{1}{2}\rho U_r^2 C_Y(\psi)Ah_v \end{cases} \quad (6\text{-}18)$$

where F_{wx}, F_{wy}, F_{wz}, $M_{w\phi}$, $M_{w\theta}$, and M_{wz} are the drag force, side force, lift force, rolling moment, pitching moment and yawing moment acting on the vehicle, respectively. C_D, C_S, C_L, C_R, C_P and C_Y are the coefficients of drag force, side force, lift force, rolling moment, pitching moment and yawing moment for the vehicle, respectively. "A" is the frontal area of the vehicle; h_v is the distance from the center of gravity of the vehicle to the road surface; U_r is the relative wind velocity to the vehicle, which is defined as:

$$U_r^2(x,t) = \left[V + (U + u(x,t))\cos\beta\right]^2 + \left[(U + u(x,t))\sin\beta\right]^2$$

$$\tan\psi = \frac{(U + u(x,t))\sin\beta}{V + (U + u(x,t))\cos\beta} \tag{6-19}$$

where V is the driving speed of vehicle; U and $u(x,t)$ are the mean wind velocity and turbulent wind velocity component on the vehicle, respectively; β is the attack angle of the wind to the vehicle, which is the angle between the wind direction and the direction in which the vehicle is moving; and ψ is usually between 0 and π.

The time history of the turbulent wind velocity component $u(t)$ and $w(t)$ can be generated using fast spectral representation method proposed by Cao et al. (2000). The time history of wind component $u(t)$, at the jth point along the bridge span can be generated with (Cai and Chen 2004):

$$u_j(t) = \sqrt{2(\Delta\omega)}\sum_{m=1}^{j}\sum_{q=1}^{N_f}\sqrt{S(\omega_{mq})}G_{jm}(\omega_{mq})\cos(\omega_{mq}t + \psi_{mq}), j = 1,2,\cdots,N_s \tag{6-20}$$

where N_f is a sufficiently large number representing the total number of frequency intervals; N_s is the total number of points along the bridge span to simulate; S is the spectral density of turbulence in along-wind direction (Kaimal spectrum for $u(t)$ and Panofsky-McCormick spectrum for $w(t)$); ψ_{mq} is a random variable uniformly distributed between 0 and 2π; $\Delta\omega = \omega_{up}/N_f$ is the frequency increment; ω_{up} is the upper cutoff frequency with the condition that the value of $S(\omega)$ is less than a present small number ε when $\omega > \omega_{up}$ and

$$G_{jm}(\omega) = \begin{cases} 0 & \text{when } 1 \le j - m \le Ns, \\ C^{|j-m|} & \text{when } m = 1, m \le j \le Ns, \\ C^{|j-m|}\sqrt{1-C^2} & \text{when } 2 \le m \le j \le Ns. \end{cases} \quad ; \quad C = \exp\left(\frac{-7\omega\Delta}{2\pi U}\right) \tag{6-21}$$

where Δ is the distance between two consecutive simulated points.

6.3.4. Equivalent Nodal Force

In terms of the finite element method, the interaction force between the vehicle tire and bridge deck may not apply at element node as the vehicle passes over the bridge. Therefore, the interaction force, i.e. $\{F_b^c\}$ in Eq. (6-6) needs to be transformed to equivalent nodal force $\{F_b^{cN}\}$ in the analysis. Nevertheless, the wind forces in Eqs. (6-14) to (6-17) acting on the center of elasticity of the deck cross section need to be distributed to the nodes of the deck section.

According to the virtual work principle, the works done by an equivalent nodal force and an actual force should be equal:

$$\{d_{b\_nodal}\}^T\{F_b^N\} = \{d_{b\_contact}\}^T\{F_b\} = ([N_b]\{d_{b\_nodal}\})^T\{F_b\} = \{d_{b\_nodal}\}^T([N_b]^T\{F_b\}) \tag{6-22}$$

$$\{F_b^N\} = [N_b]^T\{F_b\} \tag{6-23}$$

where $\{d_{b\_nodal}\}$ is the bridge deck nodal displacement, $\{d_{b\_contact}\}$ is the displacement of bridge-vehicle contact points, and $[N_b]$ is the shape function of the bridge deck element.

Similarly, the wind forces acting on the center of elasticity of the deck cross section are distributed to the nodes of the deck section either in terms of wind pressure distribution around the deck section (Xu et al 2009) or by applying the virtual work principle (Chen et al. 2011a). By applying the virtual work principle, the wind forces at the center of elasticity of the ith section can be distributed to all nodes (Chen 2010):

$$\{F_b^{wn}\}^i = [N_b^w]\{F_b^w\}^e \tag{6-24}$$

where $[N_b^w]^i$ is the displacement transformation matrix, $\{F_b^w\}^e$ is the wind forces at the section of the bridge deck, and $\{F_b^{wn}\}^i$ is the wind forces at the nodes of the section.

6.3.5. Numerical Solutions to the Coupled Equations

To simplify the modeling procedure in the bridge-vehicle coupled system, the bridge mode superposition technique is used. The bridge mode shape $\{\Phi_i\}$ and the corresponding natural circular frequencies ω_i are firstly obtained from bridge modal analysis by using conventional finite element software such as ANSYS. The bridge dynamic response $\{d_b\}$ can be expressed as:

$$\{d_b\} = \left[\{\Phi_1\}\{\Phi_2\}\ldots\{\Phi_n\}\right]\{\xi_1\ \xi_2\ \ldots\xi_n\}^T = [\Phi_b]\{\xi\} \tag{6-25}$$

where n is the total number of modes for the bridge under consideration; and $\{\Phi_i\}$ and ξ_i are the ith mode shape and its generalized coordinates, respectively. Each mode shape is normalized such that $\{\Phi_i\}^T[M_b]\{\Phi_i\} = 1$ and $\{\Phi_i\}^T[K_b]\{\Phi_i\} = \omega_i^2$. The damping matrix $[C_b]$ is assumed to be $2\omega_i\eta_i[M_b]$, where ω_i denotes the natural circular frequency of the bridge and η_i is the percentage of the critical damping for the bridge's i^{th} mode. Eq. (6-6) can be rewritten as:

$$[I]\{\ddot{\xi}_b\} + [2\omega_i\eta_i I]\{\dot{\xi}_b\} + [\omega_i^2 I]\{\xi_b\} = [\Phi_b]\{F_b^{cN}\} + [\Phi_b]\{F_b^{wN}\} \tag{6-26}$$

The mode superposition approach makes it possible to separate the bridge modal analysis from vehicle-bridge coupled model. Consequently, the coupled vehicle-bridge system vectors contain the modal components of the bridge rather than its physical components, and the physical components of the vehicles. The degrees of freedom, the number of equations in Eq. (6-6), and the complexity of the whole procedure are greatly reduced.

After transforming the contact forces and wind forces into equivalent nodal forces and substituting them into Eqs. (6-6) and (6-7), the final equations of motion for the coupled system are as follows:

$$\begin{bmatrix} M_b & \\ & M_v \end{bmatrix}\begin{Bmatrix} \ddot{d}_b \\ \ddot{d}_v \end{Bmatrix} + \begin{bmatrix} C_b + C_{bb} + C_{bw}^{se} & C_{bv} \\ C_{vb} & C_v \end{bmatrix}\begin{Bmatrix} \dot{d}_b \\ \dot{d}_v \end{Bmatrix} + \begin{bmatrix} K_b + K_{bb} + K_{bw}^{se} & K_{bv} \\ K_{vb} & K_v \end{bmatrix}\begin{Bmatrix} d_b \\ d_v \end{Bmatrix} = \begin{Bmatrix} F_{bc} + F_{bw}^{st} + F_{bw}^{b} \\ F_{vc} + F_v^G + F_{vw} \end{Bmatrix}$$
(6-27)

The additional terms C_{bb}, C_{bv}, C_{vb}, K_{bb}, K_{bv}, K_{vb}, F_{bc}, F_{vc}, C_{bw}^{se}, K_{bw}^{se}, F_{bw}^{st} and F_{bw}^{b} in Eq. (6-27) are due to the expansion of the contact force in comparison with Eqs. (6-6) and (6-7). When the vehicle is moving along the bridge, the bridge-vehicle contact points change with the vehicle position and the road roughness at the contact point. Consequently, the contact force between the bridge and vehicle changes, indicating that the addition terms in Eq. (6-27), for instance, C_{bb}, C_{bv}, C_{vb}, K_{bb}, K_{bv}, K_{vb}, F_{bc}, F_{vc}, C_{bw}^{se}, K_{bw}^{se}, F_{bw}^{st} and F_{bw}^{b}, are time dependent terms and will

128

change as the vehicle moves across the bridge.

The mode superposition makes it possible to separate the bridge modal analysis from vehicle-bridge coupled model. Then Eq. (6-27) changes to:

$$\begin{bmatrix} I & \\ & M_v \end{bmatrix} \begin{Bmatrix} \ddot{\xi}_b \\ \ddot{d}_v \end{Bmatrix} + \begin{bmatrix} 2\omega_i \eta_i I + \Phi_b^T C_{bb} \Phi_b + \Phi_b^T C_{bw}^{se} \Phi_b & \Phi_b^T C_{bv} \\ C_{vb} \Phi_b & C_v \end{bmatrix} \begin{Bmatrix} \dot{\xi}_b \\ \dot{d}_v \end{Bmatrix}$$
$$+ \begin{bmatrix} \omega_i^2 I + \Phi_b^T K_{bb} \Phi_b + \Phi_b^T K_{bw}^{se} \Phi_b & \Phi_b^T K_{bv} \\ K_{vb} \Phi_b & K_v \end{bmatrix} \begin{Bmatrix} \xi_b \\ d_v \end{Bmatrix} = \begin{Bmatrix} \Phi_b^T F_{bc} + \Phi_b^T F_{bw}^{st} + \Phi_b^T F_{bw}^{b} \\ F_{vc} + F_v^{G} + F_{vw} \end{Bmatrix} \tag{6-28}$$

The coupled vehicle-bridge-wind system vectors contain modal components of the bridge and the physical components of the vehicles. Consequently, the number of equations in Eq. (6-27) and the complexity of the whole procedure are greatly reduced.

Eq. (6-28) is solved by the Rouge-Kutta method in time domain. At each time step, the contact force at each contact point is calculated. If this force is in tension, which means the corresponding vehicle tire leaves the riding surface, then the force at this contact point is set to zero and the corresponding time dependent terms in Eq. (6-28) are also modified. In this model, the vehicle can jump or leave the riding surface, i.e., the vehicle tires are not necessary to remain in contact with the bridge deck at all time. After obtaining the bridge dynamic response $\{d_b\}$, the stress vector can be obtained by:

$$[S] = [E][B]\{d_b\} \tag{6-29}$$

where $[E]$ is the stress-strain relationship matrix and is assumed to be constant over the element in bridge's life cycle and $[B]$ is the strain-displacement relationship matrix assembled with x, y and z derivatives of the element shape functions. Finally, the stress ranges for bridge details can be obtained for a given vehicle speed, mean wind velocity and road roughness condition.

6.4 Acquisition of Stress Cycle Blocks

6.4.1. Definition of Stress Cycle Blocks

The wind induced vibrations for the whole bridge are in kilo-meter scale; while the vehicle induced local dynamic impacts are within limited influence areas in meter scale (Chan et al. 2008). As a result, the repeated block of cycles for wind loads and vehicle loads are different. It has been verified that the strain history of bridges under normal traffic can be approximately represented by a repeated daily block of cycles (Li et al 2002). For the wind induced dynamic effects, such cycles are hourly repeated (Chen et al 2011b). For each block of stress cycles, the stress history in the block varies with the random variables used in the vehicle-bridge-wind dynamic system. These parameters, for instance, the road roughness coefficient, vehicle speed distribution, and wind velocity distribution, can be obtained from the road condition assessment, the traffic information, or the meteorological data. Therefore, the dynamic stress histories in each block of stress cycles are able to be generated randomly based on the results from numerical simulations.

6.4.2. Progressive Road Roughness Deterioration Model

In order to include the progressive pavement damages due to traffic loads and environmental corrosions, a progressive road roughness deterioration model for the bridge deck surface is used (Zhang and Cai 2011):

$$\phi_t(n_0) = 6.1972 \times 10^{-9} \times \exp\left\{\left[8.39 \times 10^{-6}\, \phi_0 e^{\eta t} + 263(1 + SNC)^{-5} (CESAL)_t\right] / 0.42808\right\} + 2 \times 10^{-6} \quad (6\text{-}30)$$

where ϕ_t is the road roughness coefficient at time t; ϕ_0 is the initial road roughness coefficient directly after completing the construction and before opening to traffic; t is the time in years; η is the environmental coefficient varying from 0.01 to 0.7 depending upon the dry or wet, freezing or non-freezing conditions; SNC is the structural number modified by sub grade strength and $(CESAL)_t$ is the estimated amount of traffic in terms of AASHTO 18-kip cumulative equivalent single axle load at time t in millions.

Five road roughness classifications are defined by the International Organization for Standardization (1995), and the ranges for the road roughness coefficients (RRC) are listed in Table 6-2. The road roughness coefficient for the current block of stress cycles is calculated based on the traffic information or can be adopted from the measured RRC records for existing bridges. In order to save calculation cost, the calculated or measured RRC is classified into one of the five classifications for the vehicle-bridge dynamic analysis. If the RRC exceeds the maximum values for the very poor conditions (2.048×10^{-3}), a surface renovation is expected. If that is the case, the road surface condition is re-assessed and the road roughness condition will most likely be "very good" and deteriorate again as time goes.

Table 6-2   RRC values for road roughness classifications

Road roughness classifications	Ranges for RRCs
Very good	2×10^{-6} - 8×10^{-6}
Good	8×10^{-6} - 32×10^{-6}
Average	32×10^{-6} - 128×10^{-6}
Poor	128×10^{-6} - 512×10^{-6}
Very poor	512×10^{-6} - 2048×10^{-6}

6.4.3.   Vehicle Types and Speeds

Traffic loads have been traditionally evaluated with the data from weight-in-motion (WIM) or traffic spectrum collected from the site (Oh et al. 2007). The common practice in the analysis of the interactions of long span bridges and vehicles is to choose only one vehicle or a few identical vehicles in one line (Guo and Xu 2001, Chen and Cai 2004). The position of vehicles and interval settings are pre-decided based on considerations to simulate the normal traffic condition and accumulate normal fatigue damages due to vehicles. In the present study, a HS20-44 truck is used as the prototype of the vehicle. The geometry, mass distribution, damping, and stiffness of the tires and suspension systems of the truck can be found in the previous studies (Zhang and Cai 2011).

The dynamic displacements and stress ranges of bridge details were found to be changing with the vehicle speed based on previous studies (Cai and Chen 2004; Cai et al. 2007). Typically, the maximum speed limits posted to bridges or roads are based on the 85th percentile speed when adequate speed samples are available. The 85th percentile speed is a value that is used by many states and cities for establishing regulatory speed zones (Donnell et al. 2009; TxDOT 2006). Statistical techniques show that a normal distribution occurs when random samples of traffic are measured (TxDOT 2006). This allows describing the vehicle speed conveniently with two characteristics, i.e. the mean and standard deviation. In the present study, the 85th percentile speed is approximated as the sum of the mean value and one standard deviation for simplification. The speed limit is assumed in the present study as 31.3m/s (70mph) and the coefficient of variance of vehicle speeds is assumed as 0.2, which leads to a mean vehicle speed of 26.1 m/s (58.3mph). In order to simplify the calculations, the randomly generated vehicle speeds are grouped into six ranges that are represented by the vehicle speed from 10m/s (22.4 mph) to 60m/s (134.4mph). The probabilities of vehicle speed in all ranges are calculated based on normal distribution and listed in Table 6-3.

Table 6-3 Vehicle speed ranges

U_{ve}	Vehicle speed range	Probability
10m/s (22.4mph)	<15m/s (33.6 mph)	5.62E-02
20m/s (44.8mph)	15m/s (33.6mph) - 25m/s (56mph)	4.44E-01
30m/s (67.2mph)	25m/s (56mph) -35m/s(78.4 mph)	4.44E-01
40m/s (89.6mph)	35m/s (78.4 mph) - 45m/s (100.8 mph)	5.55E-02
50m/s (112mph)	45m/s (100.8 mph) - 55m/s (123.2 mph)	7.50E-04
60m/s (134.4 mph)	>55m/s (123.2 mph)	9.59E-06

6.4.4. Traffic Simulation

Transportation Research Board classifies the Level of Service (LOS) from A to F based on the range of the traffic occupancy in the Highway Capacity Manual (TRB 2000). Three representative traffic conditions and their corresponding occupancies "r" are considered in the present study. "Free flow" with $r = 0.07$ corresponds to Level-of-Service A~B (9veh/km/lane). In this case the traffic can flow at or above the posted speed limit and effects of incidents or point breakdowns are easily absorbed. "Moderate flow" with $r = 0.15$ corresponds to Level-of-Service C~D (20veh/km/lane). In this case a near free-flow or decreasing free-flow operations are maintained and the traffic speeds slightly decrease as the traffic volume slightly increases. In this flow condition, freedom to maneuver with the traffic stream is much more limited and driver comfort levels decrease. "Busy flow" with $r = 0.24$ corresponds to Level-of-Service E~F (32veh/km/lane). In this case traffic flow becomes irregular or forced to stop and the speed varies rapidly with frequent slowing required. In the present study, one day is considered to calculate the numbers and magnitude of the stress ranges. Therefore, the total number of vehicles N_v in the one-day period for a given vehicle speed v m/s are obtained and listed in Table 6-4.

Table 6-4 Number of vehicles passing the bridge detail in one day

Occupancy	Free flow	Moderate flow	Busy flow
r	0.07	0.15	0.24
Level-of-Service	A~B (9veh/km/lane)	C~D (20veh/km/lane)	E~F (32veh/km/lane)
Numbers of vehicles in one day	777.6v	1728v	2764.8v

Traffic conditions vary in different months in a year and in different hours in a day. Based on generic variation data from National Research Council, Wu (2010) summarized the percentages of the total hours for each representative traffic condition in each month as shown in Fig. 6-2 (a). The busy flow occurs very rare (<0.5%) compared to the other two categories and the average probability occurrence of free flow and moderate flow is 74% and 26%, respectively. Traffic volumes also vary with hours. Since several hours in a typical day have only small percentage of traffic, those hours are considered to be condensed and a total of 14 hours are considered to have a certain traffic condition classified by LOS while the other 10 hours are considered as no traffic condition. Therefore, the normal traffic condition is defined as the combination of no traffic condition, free low condition and moderate flow condition. As a result, the probability of occurrence of no traffic flow, free flow and moderate flow condition can be obtained as 10/24=41.7%, 14/24×74%=43.2% and 14/24×26%=15.2%, respectively. In order to include the effects of the busy flow condition, which will not last for one hour in a day, the probability of occurrence of busy traffic flow condition is obtained as 14/24×0.5%=0.3%.

Based on a preliminary study, the responses of the bridge details decrease drastically when the vehicle are far away from the bridge details. As a result, the dynamic stress ranges from a vehicle do not overlap with that from the following vehicle. The dynamic stress range for a given bridge detail is

131

superimposed by the dynamic stress ranges from wind loads and individual vehicles with varied speeds and different road roughness conditions.

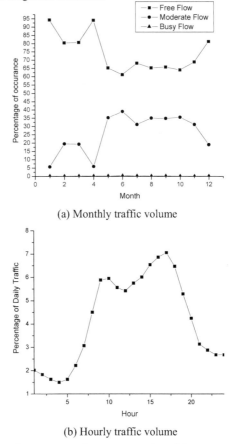

(a) Monthly traffic volume

(b) Hourly traffic volume

Fig. 6-2 Percentage of occurrence (adapted from Wu 2010)

6.4.5. Wind Environments

Based on the wind data recorded in the Xiaoyangshan Meteorological Observatory near the bridge location, the wind velocities and directions in the area were obtained. The wind velocity at the bridge deck height can be obtained via the wind profile power law with the exponent being 0.1. Probability of the wind direction and speed in a typical year are listed in Fig. 6-3 and Table 6-5 based on the records in four continuous years (Ge et al. 2003). The wind velocity is assumed to have a Weibull distribution with the value of the shape parameter being 2, which makes this distribution a Rayleigh distribution.

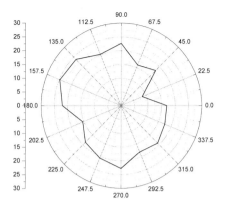

(a) Maximum wind velocity (m/s)

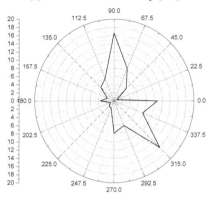

(b) Probability of wind direction (%)

Fig. 6-3 Wind roses

Nevertheless, the location of this bridge is exposed to the strong wind from tropical cyclones or typhoons. According to the recorded 129 instances of strong wind from tropical cyclones or typhoons in the 36 years from 1960 to 1995, 92% of the wind directions were north or south, which is along the bridge, and 8% of the wind directions were east or west, which might induce large structure responses. As a result, on an average basis, each year there is about 129/36×8%=0.3 instances of cross winds that might induce large vibrations.

Table 6-5 Wind velocity and directions for Donghai Bridge

Wind Direction	Yaw Angle (°)	Mean wind Velocity (m/s)	Probability (%)	Mean crosswind Velocity (m/s)
ENE	22.5	2.8	0.8	2.6
NE	45	4.5	4.3	3.2
NNE	67.5	4.6	8.6	1.8
N	90	5.1	16.6	0.0
NNW	112.5	5.7	5.9	2.2
NW	135	6.9	4.7	4.9
WNW	157.5	4.5	1.8	4.2
W	180	4.0	3.4	4.0
WSW	202.5	2.8	1.0	2.6
SW	225	3.1	1.8	2.2
SSW	247.5	3.5	1.9	1.3
S	270	4.5	7.9	0.0
SSE	292.5	5.1	6.5	2.0
SE	315	4.3	16.1	3.0
ESE	337.5	3.7	7.7	3.4
E	360	3.8	10.8	3.8

In addition, the maximum recorded wind velocities in the four seasons range from 20m/s to 25m/s. Among the large wind velocities induced by tropical cyclones, 50% of strong winds last for 6 to 12 hours and 28% and 21% last for 18-36 hours and more than 42 hours, respectively. In order to include strong wind effects in the bridge's life cycle, the strong wind velocities of 20-25m/s are assumed to last $0.3 \times (0.5 \times 9 + 0.28 \times 27 + 0.21 \times 42) = 6$ hours in one year. As a result, the probability of wind velocity in a typical year can be obtained based on the distribution of wind velocities and directions with the superposition of the probability of strong wind induced by tropical cyclones or typhoons. Five mean wind velocities are chosen to represent the wind velocity ranges in the bridge location with a calculated probability as listed in Table 6-6. It is noteworthy that the probability of wind velocity exceeding 20m/s was calculated based on the 6 hours' strong wind due to tropical cyclones or typhoons. In addition, except explicitly specified, the considered wind attack angle of the bridge is zero.

Table 6-6 Representative mean wind velocity and probability

U_{ve}	Ranges	Probability (%)
0 m/s	<0.5m/s	28.8
2m/s (4.5mph)	0.5m/s-3.5m/s	60.2
5m/s (11.2mph)	3.5m/s-6.5m/s	10.4
8m/s (18.0mph)	6.5m/s-11.5m/s	0.54
15m/s (33.6mph)	11.5m/s-20m/s	0.001
25m/s (56.0 mph)	>20m/s	0.068

6.5 Fatigue Reliability Assessment

6.5.1. Fatigue Damage Model

For variable amplitude stress cycles, the Palmgren-Miner damage law, which is also called as the linear fatigue damage rule (LDR), is often used (Miner 1945, Byers et al. 1997 a):

$$D(t) = \sum_i \frac{n_i}{N_i} = \frac{n_{tc}}{N} \tag{6-31}$$

where n_i is number of observations in the predefined stress-range bin S_{ri}, N_i is the number of cycles to failure corresponding to the predefined stress-range bin; n_{tc} is the total number of stress cycles and N is the number of cycles to failure under an equivalent constant amplitude loading (Kwon and Frangopol 2010):

$$N = A \cdot S_{re}^{-m} \tag{6-32}$$

where S_{re} is the equivalent stress range and A is the detail constant taken from Table 6.6.1.2.5-1 in AASHTO LRFD bridge design specifications (AASHTO 2010). Either using the Miner's rule or Linear Elastic Fracture Mechanics (LEFM) approach, the equivalent stress range for the whole design life is obtained through the following equation (Chung 2004):

$$S_{re} = \left(\sum_{i=1}^{n} \alpha_i \cdot S_{ri}^{m} \right)^{1/m} \tag{6-33}$$

where α_i is the occurrence frequency of the stress-range bin, n is the total numbers of the stress-range bin and m is the material constant that could be assumed as 3.0 for all fatigue categories (Keating and Fisher 1986).

Since each truck passage might induce multiple stress cycles, two correlated parameters are essential to calculate the fatigue damages done by each truck passage, i.e. the equivalent stress range and the number of stress cycles caused by each truck passage. After counting the number of stress cycles at different stress range levels using the rainflow counting method, fatigue damage increment ΔD_i can be obtained:

$$\Delta D_i = \sum_j \frac{n_j}{N_j} \tag{6-34}$$

6.5.2. Limit State Function

When the fatigue damage variable D increases to 1, a fatigue failure is expected. In the probabilistic approach, a limit state function (LSF) needs to be defined first in order to ensure target fatigue reliability (Nyman and Moses 1985):

$$g(X) = D_f - D_i \tag{6-35}$$

where D_f is the damage to cause failure and is treated as a random variable with a mean value of 1; and g is a failure function such that $g<0$ implies a fatigue failure. The accumulated damage at the end of stress block i is

$$D_i = D_{i-1} + \Delta D_i \tag{6-36}$$

and ΔD_i is the fatigue damage increment at stress block i as shown in Eq. (6-34).

Based on the information from the literature, the related random variables are listed in Table 6-7, including their distribution types, mean values, coefficients of variation (COVs) and descriptions. As a result, the fatigue damage D_i is calculated using Eq. (6-34) and (6-36). Based on the defined LSF in Eq. (6-35), a conditional probability of failure after the fatigue damage accumulation of the present block of stress cycles is obtained and recorded. The fatigue reliability for a given design life of a bridge can be obtained. The total accumulated probability of failure due to all of the preceding blocks of stress cycles can be calculated and compared with the maximum allowable value of probability of failure corresponding to the target reliability index. If the accumulated probability of failure is less than the maximum allowable value, the analysis will continue to the next block of stress cycles. Otherwise, the cycle will stop and the fatigue life for the target reliability index can be obtained. As a result, the fatigue reliability index can be obtained based on Eq. (6-35). In the present study, the target reliability index β is chosen as 3.5, which is typically used in AASHTO LRFD (2010).

Table 6-7 Summary of LSF parameters

Parameter	Mean	COV	Distribution	Description
D_f	1.0	0.15	Lognormal	Damage to cause failure
A	7.83×10^{10}	0.34	Lognormal	Detail constant
M	3.0		Deterministic	Slope constant
V	55.8mph (25.0m/s)	0.2	Normal	Vehicle speed

6.6 Selected Results

6.6.1. Stress Ranges due to Vehicle Loads

For long-span bridges, the superimposed dynamic stress ranges are generated from wind induced vibrations and static or dynamic effects from vehicles. Due to the large differences of the natural frequencies of the vehicle and the bridge, the stress cycle is one for each truck passage and the dynamic amplification factor is relatively small for long-span suspension bridges (Chen et al. 2011b). Similar results can also be found in the present case study. Therefore, the numbers of stress ranges from vehicles are equal to the numbers of truck passages during the calculation period of one day as listed in Table 6-4. The stress ranges from a HS20-44 truck at different vehicle speeds and road roughness conditions are listed in Table 6-8. It is noteworthy that the stress range from static vehicle loads are 9.1 MPa. The dynamic stress ranges from vehicle loads vary little with each other for different vehicle speeds and road roughness conditions. The dynamic effect from vehicle loads ranges from 0.5 MPa to 1.6 MPa. That is 5% to 18% of the stress ranges from the static vehicle load. Since the dynamic stress range varies little for different vehicle speeds, the mean value of the stress ranges, 9.9 MPa, is assumed to represent the stress ranges for all the vehicle speeds and road roughness conditions.

Table 6-8 Stress ranges due to single vehicle passage

U_{ve} \ Roughness	very good (MPa)	good (MPa)	Average (MPa)	poor (MPa)	very poor (MPa)	Mean value (MPa)
10m/s (22.4mph)	9.7	9.8	9.7	9.7	10.0	9.8
20m/s (44.8mph)	10.0	9.7	9.7	10.0	10.5	10.0
30m/s (67.2mph)	9.7	9.6	9.7	9.7	10.2	9.8
40m/s (89.6mph)	9.8	9.8	9.8	10.1	9.8	9.9
50m/s (112mph)	10.0	9.8	9.7	10.1	10.3	10.0
60m/s (134.4 mph)	9.6	9.9	10.0	10.4	10.7	10.1
Mean value	9.8	9.8	9.8	10.0	10.3	9.9

6.6.2. Stress Ranges due to Wind Effects

In each block of stress cycles, cycle counting methods, such as the rainflow counting method, are used to obtain the stress range values and the number of cycles from wind for a given wind velocity. Since the stress range cut-off levels change the number of cycles greatly, a reasonable value is necessary. In the data analysis of stress ranges obtained from field monitoring, 3.45 MPa (0.5ksi) is a typical cut-off level for stress ranges to calculate the numbers per truck passage. A similar cut-off level from 3.45 MPa (0.5ksi) to 33% of the constant amplitude fatigue limit (CAFL) was suggested by Kwon and Frangopol (2010). Since the contribution of stress ranges less than 3.45 MPa (0.5ksi) can be neglected, the cut-off level of the stress range of 3.45 MPa (0.5ksi) is chosen in the present study.

Since the longitudinal wind has little dynamic effects, only the dynamic effects from the component of the crosswind are considered. The mean values for the stress ranges and the numbers of stress cycles for one hour are listed in Table 6-9, which indicates that the equivalent stress range and the number of stress cycles increase with the increase of wind velocity.

Table 6-9 Stress ranges and number of stress cycles due to wind loads in one hour

U_{ve}	Equivalent Stress range values (MPa)	Numbers of stress cycles
2m/s (4.5mph)	5.5	5,628
5m/s (11.2mph)	11.8	15,479
8m/s (18.0mph)	18.2	12,665
15m/s (33.6mph)	22.5	37,759
25m/s (56.0 mph)	38.0	49,016

6.6.3. Combined Dynamic Load Effects

The size of the stress cycle blocks might vary with each other, for instance, the stress cycle block is assumed lasting for one hour for wind loads and one day for vehicle loads (Li et al 2002, Chen et al 2011b). In the present study, one day is used to calculate the dynamic stress ranges from wind and vehicles. For each stress block, dynamic stress ranges from wind and vehicles are superposed together and the rainflow counting method is used to obtain the stress range values and the number of cycles from wind and vehicles.

According to the previous studies (Xu and Guo 2003, Cai and Chen 2004), heavy trucks are critical to bridge dynamic behaviors and the dynamic effects from light trucks or sedan are much smaller. In the present study, only the dynamic load effects from the 3-axles trucks are presented. When the speed limit is 31.3m/s (70mph) and the coefficient of variance is 0.2, the mean vehicle speed is 25m/s (56mph). If 10% of all vehicles are 3-axles trucks, the average vehicle numbers for free flow, moderate flow, and busy flow are 1944, 4320, and 6912 based on Table 4. The stress ranges and numbers of cycles due to the dynamic loads from wind and vehicles are listed in Table 10 for an average road roughness condition.

Table 6-10 Stress ranges and stress cycles due to dynamic loads in one day

Traffic \ Wind velocity	No traffic		Free flow		Moderate flow		Busy flow	
	(MPa)	Number	(MPa)	Number	(MPa)	Number	(MPa)	Number
No wind	---	--	9.8	19,459	9.8	43,200	9.8	69,120
2 m/s	5.5	5,628	6.6	129,950	8.5	170,030	8.5	272,000
5 m/s	11.8	15,479	11.1	374,300	11.7	350,700	12.5	446,600
8 m/s	18.2	12,665	15.1	342,250	17.3	360,260	18.5	465,600
15 m/s	22.5	37,759	20.3	899,110	21.2	1,088,300	24.0	1,643,500
25 m/s	38.0	49,016	29.2	1,175,300	30.1	903,700	31.3	1,140,600

6.6.4. Fatigue Life Estimations

For comparison, the fatigue life based on a reliability index of 3.5 obtained through the developed procedure with different combinations of wind velocity and traffic flow information is listed in Table 6-11.

137

Table 6-11 Estimated fatigue life (year)

Wind Velocity / Traffic	No Traffic	Free Flow	Moderate Flow	Busy Flow	Normal Traffic Distribution
No wind	---	317	146	93	410
2 m/s	5660	159	59	37	183
5 m/s	230	13	12	8	20
8 m/s	80	6	4	3	8
Wind velocity distribution at bridge site	1024	57	39	25	83

There are five categories in the table representing different traffic conditions including no traffic condition, free flow condition, moderate flow condition, busy flow condition and the normal traffic condition at the bridge site. In the normal traffic condition, as discussed earlier, the probability of occurrence of no traffic flow, free flow, moderate flow and busy flow condition is 41.7%, 43.2%, 15.2% and 0.3%, respectively.

The wind velocity consists of the mean wind velocity and the turbulence wind velocity. In the table, the listed wind velocities of 2m/s to 8m/s are the mean crosswind velocity. The turbulence wind velocities are simulated along the bridge span length based on Eq. (20). The total number of frequency intervals Nf equals to 1024 and the upper cutoff frequency equals to 2π. The given wind, for example 8m/s is assumed to be acting on the bridge site all the time, namely 100% occurrence. In comparison, for the actual wind velocity distribution at the bridge site, as discussed earlier and listed in Table 6, the probability of occurrence of representing mean crosswind velocity of 0 m/s, 2m/s, 5m/s, 8 m/s, 15 m/s and 25m/s is 28.8%, 60.2%, 10.4, 0.54%, 0.001% and 0.068%, respectively.

Three cases, for instance, traffic only cases, wind loads only and combined loads of traffic and wind, are discussed here:

(a) Traffic loads only. If only traffic loads are considered without wind effects, the fatigue life for different traffic flow conditions varies from 93 years to 317 years for the free to busy flow conditions and 410 years for the normal traffic condition which is longer than the free flow condition.

(b) Wind loads only. When only wind loads are considered without traffic loads, the fatigue life varies from 80 years to 5660 years for the mean crosswind velocity ranging from 2m/s to 8m/s. The corresponding fatigue life considering the bridge wind condition is 1024 years.

(c) Wind and traffic combined loads. When both the traffic and wind effects are included, the fatigue life drops accordingly. The fatigue life for the varied traffic conditions from the free to busy flow for the mean crosswind velocity of 2m/s, 5m/s, and 8 m/s varies from 37 to 159 years, 8 to 13 years, and 3 to 6 years, respectively. When the wind velocity distribution at the bridge site is considered, the estimated fatigue life ranges from 25 years to 57 years for traffic flow condition varying from the busy flow condition to the free flow condition. If the normal traffic distribution at the bridge site is considered, the fatigue life ranges from 8 years to 183 years when the mean crosswind velocity ranges from 2m/s to 8m/s. If both of the normal traffic condition and wind velocity distribution at the bridge site are considered, the fatigue life is 83 years, which is slightly less than the design life of 100 years of the bridge. The combination effects from the wind loads and vehicle loads decrease the fatigue life drastically. In an extreme condition, for instance, when the mean crosswind velocity is 8m/s for all the time and the traffic flow is always busy in the bridge's life cycle, the fatigue life for the bridge with the target reliability index of 3.5 is only 3 years.

In summary, the vehicle loads or wind loads alone might not be able to induce serious fatigue problems for details of long-span bridges under normal traffic condition or normal wind velocities. However, if the combined dynamic effects from wind and vehicles are considered, fatigue damage accumulation might endanger the bridge's safety and reliability. Generally, the fatigue life decrease with the increase of traffic volume and wind velocities.

6.7 Concluding Remarks

This paper presents a comprehensive framework for fatigue reliability analysis for long-span bridges under combined dynamic loads from vehicles and winds. After modeling the long-span bridges with multiple complicated structural details with equivalent orthotropic material shell elements, dynamic stress ranges of bridge details are obtained via solving the equations of motions for the vehicle-bridge-wind dynamic system with multiple random variables considered, for instance, vehicle speeds, road roughness conditions, and wind velocities and directions. Therefore, the dynamic stress histories in each block of stress cycles are randomly generated. After counting the number of stress cycles at different stress range levels using the rainflow counting method, the fatigue damage increment can be obtained using the fatigue damage accumulation rule. The probability of failure for the fatigue damage at the end of each block of stress cycles and the cumulative probability of failure can be obtained. Therefore, the progressive fatigue damage accumulation in the bridge's life cycle is achieved and the fatigue life and reliability for the given structural details is obtained. From the present study, the following conclusions are drawn:

1. The dynamic effects from vehicles are relatively small for long-span bridges and the effects from vehicle speeds and road roughness conditions can be neglected.

2. The dynamic stress ranges and numbers of cycles increase with the increase of wind velocity.

3. The combined dynamic effects from winds and vehicles might result in serous fatigue problems for long-span bridges, while the traffic or wind loads alone are not able to induce serious fatigue problems.

The present study has demonstrated an effective framework for fatigue reliability assessment of long-span bridges considering the combined dynamic effects from vehicles and wind loads. Effects from multiple random variables for the vehicle-bridge-wind dynamic system can be included. When the dynamic stresses from winds and vehicles for long-span bridges are superposed, large fatigue damage accumulations can be found, and this might endanger the bridge safety and reliability.

6.8 References

American Association of State Highway and Transportation Officials (AASHTO). (2010). LRFD bridge design specifications, Washington, DC.

Baker, C. J. (1994). "The quantification of accident risk for road vehicles in cross winds." *Journal of Wind Engineering and Industrial Aerodynamics*, 52(0), 93-107.

Byers, W. G., Marley, M. J., Mohammadi, J., Nielsen, R. J., and Sarkani, S. (1997a). "Fatigue Reliability Reassessment Procedures: State-of-the-Art Paper." *Journal of Structural Engineering*, 123(3), 271-276.

Byers, W. G., Marley, M. J., Mohammadi, J., Nielsen, R. J., and Sarkani, S. (1997b). "Fatigue Reliability Reassessment Applications: State-of-the-Art Paper." *Journal of Structural Engineering*, 123(3), 277-285.

Cai, C. S., and Chen, S. R. (2004). "Framework of vehicle-bridge-wind dynamic analysis." *Journal of Wind Engineering and Industrial Aerodynamics*, 92(7-8), 579-607.

Cai, C. S., Shi, X. M., Araujo, M., and Chen, S. R. (2007). "Effect of approach span condition on vehicle-induced dynamic response of slab-on-girder road bridges." *Engineering Structures*, 29(12), 3210-3226.

Cao, Y. H., Xiang, H. F., and Zhou, Y. (2000). "Simulation of stochastic wind velocity field on long-span bridges." Journal of Engineering Mechanics, 126(1), 1-6.

Chan, T. H. T., Li, Z. X., and Ko, J. M. (2001). "Fatigue analysis and life prediction of bridges with structural health monitoring data - Part II: Application." *International Journal of Fatigue*, 23(1), 55-64.

Chan, T. H. T., Yu, Y., Wong, K. Y., & Li, Z. X. (2008). Condition-assessment-based finite element modeling of long-span bridge by mixed dimensional coupling method. (J. Gao, J. Lee, J. Ni, L. Ma, & J. Mathew, Eds.)Mechanics of Materials. Springer. Retrieved from http://eprints.qut.edu.au/16720/

Chen, S. R., and Wu, J. (2010). "Dynamic Performance Simulation of Long-Span Bridge under Combined Loads of Stochastic Traffic and Wind." *Journal of Bridge Engineering*, 15(3), 219-230.

Chen, Z. W. (2010). "Fatigue and Reliability Analysis of Multiload Suspension Bridges with WASHMS," Ph.D dissertation, The Hong Kong Polytechnic University, Hong Kong.

Chen, Z. W., Xu, Y. L., Li, Q., and Wu, D. J. (2011a). "Dynamic Stress Analysis of Long Suspension Bridges under Wind, Railway, and Highway Loadings." *J. Bridge Eng.*, 16(3), 383.

Chen, Z. W., Xu, Y. L., Xia, Y., Li, Q., and Wong, K. Y. (2011b). "Fatigue analysis of long-span suspension bridges under multiple loading: Case study." *Engineering Structures*, 33(12), 3246-3256.

Chung, H. (2004). "Fatigue reliability and optimal inspection strategies for steel bridges," Ph.D Dissertation, The University of Texas at Austin.

Deng, L., and Cai, C. S. (2010). "Development of dynamic impact factor for performance evaluation of existing multi-girder concrete bridges." *Engineering Structures*, 32(1), 21-31.

Dodds, C. J., and Robson, J. D. (1973). "The Description of Road Surface Roughness." *Journal of Sound and Vibration*, 31(2), 175-183.

Donnell, E. T., Hines, S. C., Mahoney, K. M., Porter, R. J., and McGee, H. (2009). "Speed Concepts: Informational Guide." U.S. Department of Transportation and Federal Highway Administration. Publication No. FHWA-SA-10-001.

Ge, Y., Yang, Y. and Zhang, Z. (2003) "Report on the Wind Tunnel Experiments of Donghai Cable-stayed Bridge." Department of Bridge Engineering, Tongji University, Shanghai.

Gu, M., Xu, Y. L., Chen, L. Z., and Xiang, H. F. (1999). "Fatigue life estimation of steel girder of Yangpu cable-stayed Bridge due to buffeting." *Journal of Wind Engineering and Industrial Aerodynamics*, 80(3), 383-400.

Guo, W. H., and Xu, Y. L. (2001). "Fully computerized approach to study cable-stayed bridge-vehicle interaction." *Journal of Sound and Vibration*, 248(4), 745-761.

Honda, H., Kajikawa, Y., and Kobori, T. (1982). "Spectra of Road Surface Roughness on Bridges." *Journal of the Structural Division*, 108(ST-9), 1956-1966.

International Standard Organization. (1995). "Mechanical vibration - Road surface profiles - Reporting of measured data." Geneva.

Keating, P. B., and Fisher, J. W. (1986). "Evaluation of Fatigue Tests and Design Criteria on Welded Details." NCHRP Report 286, Transportation Research Board, Washington, D.C.

Kwon, K., and Frangopol, D. M. (2010). "Bridge fatigue reliability assessment using probability density functions of equivalent stress range based on field monitoring data." *International Journal of Fatigue*, 32(8), 1221-1232.

Laman, J. A., and Nowak, A. S. (1996). "Fatigue-Load Models for Girder Bridges." *Journal of Structural Engineering*, 122(7), 726-733.

Li, Z. X., Chan, T. H. T., and Ko, J. M. (2002). "Evaluation of typhoon induced fatigue damage for Tsing Ma Bridge." *Engineering Structures*, 24(8), 1035-1047.

Miner, M. A. (1945). "Cumulative damage in fatigue." *Journal of Applied Mechanics*, 67, A159-64.

Nyman, W. E., and Moses, F. (1985). "Calibration of Bridge Fatigue Design Model." *Journal of Structural Engineering*, 111(6), 1251-1266.

Oh, B. H., Lew, Y., and Choi, Y. C. (2007). "Realistic Assessment for Safety and Service Life of Reinforced Concrete Decks in Girder Bridges." *Journal of Bridge Engineering*, 12(4), 410-418.

Shi, X., Cai, C. S., and Chen, S. (2008). "Vehicle Induced Dynamic Behavior of Short-Span Slab Bridges Considering Effect of Approach Slab Condition." *Journal of Bridge Engineering*, 13(1), 83-92.

Transporation Research Board (TRB) (2000). Highway capacity manual, Washington, D.C.

Troitsky, M. S. (1987). *Orthotropic Bridges - Theory and Design, 2nd ed.*, The James F. Lincoln Arc Welding Foundation, Clevaland, OH.

TxDOT. (2006). "Procedures for Establishing Speed Zones." Texas Department of Transportation.

Virlogeux M. (1992). "Wind design and analysis for the Normandy Bridge". In: Larsen A, editor. Aerodynamics of large bridges. Rotterdam, the Netherlands: Balkema; P. 183-216.

Wang, T.-L., and Huang, D. (1992). "Computer modeling analysis in bridge evaluation." Florida Department of Transportation, Tallahassee, FL.

Wolchuk, R. (1963). Design Manual for Orthotropic Steel Plate Deck Bridges. Chicago, IL: American Institute of Steel Construction.

Wu, C., Zeng, M.G. and Dong, B. (2003) "Report on the Performance of Steel-Concrete Composite Beam of Donghai Cable-stayed Bridge." Department of Bridge Engineering, Tongji University, Shanghai.

Xu, Y. L., and Guo, W. H. (2003). "Dynamic analysis of coupled road vehicle and cable-stayed bridge systems under turbulent wind." *Engineering Structures*, 25(4), 473-486.

Xu, Y. L., Liu, T. T., and Zhang, W. S. (2009). "Buffeting-induced fatigue damage assessment of a long suspension bridge." *International Journal of Fatigue*, 31(3), 575-586.

Zhang, Z. T., and Ge, Y. J. (2003). "Static and Dynamic Analysis of Suspension Bridges Based on Orthotropic Shell Finite Element Method." Structural Engineers, 2003(4).

Zhang,W., C.S. Cai. (2011). "Fatigue Reliability Assessment for Existing Bridges Considering Vehicle and Road Surface Conditions", Journal of Bridge Engineering, doi:10.1061/ (ASCE) BE. 1943-5592.0000272

CHAPTER 7   CONCLUSIONS AND FURTHER CONSIDERATIONS

7.1 Summary and Conclusion

In the book, fatigue performance of existing bridges under dynamic loads from vehicles and winds are analyzed. A progressive fatigue reliability assessment approach is proposed and effects from multiple random variables in the vehicle-bridge-wind dynamic system can be included. The contribution of the dissertation can be roughly classified into three interrelated parts: (a) deeper insight of the vehicle-bridge-wind dynamic system; (b) better understanding of fatigue damage accumulation; and (c) more accurate fatigue design based on the dynamic analysis.

(a) Deeper insight of the vehicle-bridge-wind dynamic system (Chapter 2 and 5)

The major objective of the dissertation is to investigate the fatigue performance of existing bridges under dynamic load from vehicles and winds. During the life cycle of a bridge, the dynamic effects vary with the random traffic loads, the progressive deteriorated road surface conditions, and varying wind loads. Therefore, it is more realistic to use reliability methods and treat the input parameters as random variables for the dynamic system. Nevertheless, in order to avoid unmanageable large numbers of elements and degree of freedom involved in solving the equations of motions for long-span bridges with complicated structure details, an effective finite element modeling scheme is essential. The dynamic system needs to be investigated with a deeper insight.

In this study, an approach for fatigue reliability assessment of existing bridges considering the random effects of vehicle speeds and deteriorating road roughness conditions of bridge decks. After setting up the limit state function with several random variables (including fatigue damages to cause failure, vehicle speeds, road roughness conditions, the revised equivalent stress ranges and the constant amplitude fatigue thresholds), fatigue reliability of the structural details is attained. Both the normal and lognormal distribution is acceptable to describe the distribution of the revised equivalent stress range at each combination of road roughness condition and vehicle speed. From the present study, the following conclusions are drawn:

➤ The vehicle speed affects the fatigue reliability and fatigue life of the bridge components. In most cases, a higher vehicle speed induces a larger stress range and a larger number of cycles per truck passage. Accordingly, the fatigue reliability decreases with the increase of the vehicle speed.

➤ The road roughness condition influences the fatigue reliability of the bridge components. Generally, the more deteriorated road condition induces larger stress ranges and larger numbers of stress cycles for each truck passage, which leads to a smaller fatigue reliability index.

➤ With the increase of the traffic increase rate, the fatigue reliability drops and the fatigue life reduces significantly.

For long-span bridges, complicated structures details make it difficult to obtain numerically the stress range history of structural details. Local effects might be neglected by simple finite element models, while refined structural models might have unmanageable number of elements and nodes. To evaluate the structural performance under multi-scale dynamic loads, for instance, the wind loads in a kilo-meter scale and the vehicle loads in a meter scale, an effective FE model is essential. In this study, a multiple scale modeling and simulation scheme based on the equivalent orthotropic material method (EOMM) is presented. Bridge deck plates with multiple stiffeners are modeled with the elements using an equivalent orthotropic material and geometry. The bridges are assembled with simplified equivalent shell elements with the same position as the original shell element. Based on the comparison of results from modal, static, and vehicle-bridge dynamic analyses, the following conclusions are drawn:

➤ The EOMM method can be used to build the FE model with good precisions in vibration modes including the main vibrations modes and several local vibrations modes.

➤ The model built by the EOMM has a good precision in predicting static displacements, strains, and stresses.

➤ The dynamic stresses from the model built by the EOMM have a good precision if only the matched modes are used for the mode superposition techniques in the dynamic analysis.

(b) Better understanding of fatigue damage accumulation (Chapter 4)

During the most part of bridges' fatigue lives, the structure materials are in a linear range and micro cracks have not developed into macroscopic cracks. The fatigue life assessment of existing bridges is related to a sequence of progressive fatigue damage with only the initiations of micro cracks. Nonlinear cumulative fatigue damage theories were used to model the fatigue damage accumulation in this stage of the initiation of micro cracks. It is more appropriate to use the nonlinear continuous fatigue damage model for the fatigue analysis during a large fraction of bridges' life cycles. Nevertheless, the road roughness conditions deteriorated with each repeated block of stress cycles induced by multiple vehicle passages and the vehicle types, numbers, and distributions might change with time, as well.

This dissertation presents a progressive fatigue reliability assessment approach based on a nonlinear continuous fatigue damage model to include multiple random variables in vehicle-bridge dynamic system during the bridge's life cycle. Types and numbers of passing vehicles are recorded to calculate the road surface's progressive deterioration and road roughness coefficients are calculated for the each block of stress cycles. The fatigue damage accumulation and cumulative probability of failures are calculated and recorded for each block of stress cycles, as well. Once the threshold of road roughness coefficients is reached, the road profile is generated to the next category of the deteriorated road surface conditions or a road surface renovation is expected. The fatigue lives and fatigue damage index are obtained and compared with the results obtained from a linear fatigue damage model, as well. From the present study, the following conclusions are drawn:

➤ The proposed approach is effective to predict the progressive fatigue reliability of existing bridges. Different fatigue damage model and various random variables of the vehicle-bridge dynamic system in bridge's life cycle can be included in the proposed approach.

➤ Significant discrepancies of fatigue damage estimations from the NLCDR model and LDR model are found. The fatigue damage estimated by using LDR model is larger than that estimated by the NLCDR model in the early stage in bridges' life cycle. However, as the fatigue damage begins to accumulate, the fatigue damage increase rate of NLCDR model is much faster than the LDR model.

➤ Vehicle speeds have limited effects on the fatigue reliability and life, while the days of road surface discontinuities have a large effect on the fatigue reliability and life.

(c) Fatigue design based on the dynamic analysis (Chapter 3 and 6)

Under the dynamic loads from vehicles and winds, fatigue damage accumulations might endanger structural safety of existing bridges. With a deeper insight of the dynamic system and better understanding of the fatigue damage accumulation, sophisticated fatigue design approaches can be built based on the dynamic analysis on the vehicle-bridge or vehicle-bridge-wind system. Certain retrofitting actions can be performed based on the results from the fatigue reliability analysis, for instance, repairing the structure, replacing the structure, or changing the operation of the structure.

In the current bridge design specifications, a dynamic amplification factor (DAF) or dynamic load allowance (IM) is typically used to include dynamic effects from vehicles on bridges. The calculated live load stress ranges might not be correct due to the varying dynamic amplification effects in different regions along the bridge, different road roughness conditions, and multiple stress range cycles generated for one vehicle passage on the bridge. In the present study, a reliability based dynamic amplification factor on revised equivalent stress ranges (DAFS) for fatigue design is proposed to include the fatigue damages from multiple stress range cycles due to each vehicle passage at varying vehicle speeds under various road conditions in the bridge's life cycle. The effects of the long-term deck deterioration and various vehicle parameters, such as vehicle speeds and types, can be included in DAFS, as well. Parametric studies of DAFS are carried out to find the effects from multiple variables in the bridge's life cycle, for instance, the faulting days in each year, vehicle speed limit and its coefficient of variance, vehicle type distribution, and annual traffic increase rate. The calculated fatigue lives from the six different approaches, namely, DT-DAFS, PB-DAFS, PB-SWE, DT-DAF, PB-DAF, and PB-SWM, are compared with each other to acquire a reasonable fatigue life estimation to preserve both the simplicity and accuracy. From the present study, the following conclusions are drawn:

> DAFS is an effective measure of dynamic stress cycles that can include the effects from random variables in the vehicle-bridge dynamic system. Under the same target reliability level, a larger DAFS value corresponds to shorter fatigue lives.

> Faulting in the road surface increases the DAFS values and decreases the fatigue life. It has limited influence when the damages are repaired within 15 days for most cases in the present study.

> DAFS is sensitive to the road roughness deterioration rate in the bridge's life cycle. The effects of vehicle type, annual traffic increase rate, and some other parameters are reflected by the DAFS via the change of road roughness conditions in each road resurface period.

> Since DAF only reflects the largest stress amplitude while DAFS includes the fatigue damages from multiple stress range cycles due to each vehicle passage, DAF is less than the DAFS and leads to an overestimation of fatigue life.

Since long-span bridges often serve as the backbones of main transportation lines to support daily operation and hurricane evacuations, the structural reliability should be carefully assessed and predicted especially for the superposed multiple dynamic loads to ensure the structural safety. Many researches have been carried out on fatigue assessment of long-span bridges under wind loads only or vehicle loads only. This study presents a comprehensive framework for fatigue reliability analysis for long-span bridges under the combined dynamic loads from vehicles and winds. After modeling the long-span bridges with multiple complicated structural details with equivalent orthotropic material shell elements, the dynamic stress ranges of bridge details are obtained via solving the equations of motions for the vehicle-bridge-wind dynamic system with multiple random variables considered, for instance, vehicle speeds, road roughness conditions, and wind velocities and directions. The dynamic stress histories in each block of stress cycles are randomly generated. After counting the number of stress cycles at different stress range levels using rainflow counting method, the fatigue damage increment can be obtained using the fatigue damage accumulation rule. The probability of failures for the fatigue damage at the end of each block of stress cycles and the cumulative probability of failures can be obtained. Therefore, the progressive fatigue damage accumulation in the bridge's life cycle is achieved and the fatigue life and reliability for the given structural details is obtained. From the present study, the following conclusions are drawn:

➢ The dynamic effects from vehicles are relatively small for long-span bridges and the effects from vehicle speeds and road roughness conditions can be neglected.

➢ The dynamic stress ranges and numbers of cycles increase with the increase of wind velocity.

➢ The combined dynamic effects from winds and vehicles might result in serous fatigue problems for long-span bridges, though the traffic or wind load alone may not be able to induce serious fatigue problems.

7.2 Future Work

The writer believes that the following issues deserve further research:

➢ Depending on the consequence of failures, public sensitivity to failures, economic constraints to achieve structural safety, and the past experiences on design and constructions, the selection of target fatigue reliability level is a very difficult task. Therefore, calibration of the fatigue reliability index for the whole structural system to maintain the same safety level in the strength design and fatigue design is needed and can be carried out in the future work.

➢ In the dissertation, the time histories of wind speeds are generated based on stationary Gaussian process assumption. Under certain extreme weather conditions, wind speeds are usually higher and they are neither stationary nor Gaussian. For existing bridges, fatigue damages might have accumulated in the past due to those extreme weather conditions. It is necessary to calculate their effects on fatigue damage accumulation, especially when a nonlinear fatigue damage model is used. Past meteorological data can be used to include the damages from those extreme weather conditions.

➢ For a multi-scale dynamic system, the small error from the overall structural system might greatly affect the local stress of the structural details. Large uncertainties will be brought to the fatigue life estimation based on those structural details. It is necessary to build an appropriate data transfer scheme for the overall system and its sub-system in different length and time scales.

➢ For some structural details in bridges, both of the magnitude and directions of stresses might change as the magnitude, position, or direction of dynamic loads change. The multi-axle effect on fatigue damage accumulations needs to be considered. Nevertheless, micro-cracks might have developed in certain direction and partial or full stiffness matrix of the elements with micro-cracks might change, as well. Therefore, an acceleration of fatigue damage accumulation might be triggered due to such interactions. This is a very interesting direction for future research.

Made in the USA
San Bernardino, CA
19 November 2014